ETHEREUM

A Crash Course to Understanding Ethereum Mining Mechanisms for Beginners

(The Ultimate Beginner's Guide About Blockchain Wallet)

Marlin Sturgeon

Published by Tomas Edwards

Marlin Sturgeon

All Rights Reserved

Ethereum: A Crash Course to Understanding Ethereum Mining Mechanisms for Beginners (The Ultimate Beginner's Guide About Blockchain Wallet)

ISBN 978-1-990373-63-3

All rights reserved. No part of this guide may be reproduced in any form without permission in writing from the publisher except in the case of brief quotations embodied in critical articles or reviews.

Legal & Disclaimer

The information contained in this book is not designed to replace or take the place of any form of medicine or professional medical advice. The information in this book has been provided for educational and entertainment purposes only.

The information contained in this book has been compiled from sources deemed reliable, and it is accurate to the best of the Author's knowledge; however, the Author cannot guarantee its accuracy and validity and cannot be held liable for any errors or omissions. Changes are periodically made to this book. You must consult your doctor or get professional medical advice before using any of the suggested remedies, techniques, or information in this book.

Upon using the information contained in this book, you agree to hold harmless the Author from and against any damages, costs, and expenses, including any legal fees potentially resulting from the application of any of the information provided by this guide. This disclaimer applies to any damages or injury caused by the use and application, whether directly or indirectly, of any advice or information presented, whether for breach of contract, tort, negligence, personal injury, criminal intent, or under any other cause of action.

You agree to accept all risks of using the information presented inside this book. You need to consult a professional medical practitioner in order to ensure you are both able and healthy enough to participate in this program.

Table of Contents

INTRODUCTION 1

CHAPTER 1: WHAT IS ETHEREUM 4

CHAPTER 2: OPPORTUNITIES TO INVEST IN BITCOIN (CLASSIC AND CASH) VS. ETHEREUM 13

CHAPTER 3: THE DOUBLE-EDGED SWORD OF ETHEREUM 47

CHAPTER 4: WHAT IS ETHEREUM? 63

CHAPTER 5: TECHNOLOGY BEHIND ETHEREUM 70

CHAPTER 6: WHAT IS ETHEREUM? 74

CHAPTER 7: SMART CONTRACTS FROM ETHEREUM 82

CHAPTER 8: WHAT IS ETHEREUM? 94

CHAPTER 9: WHY ETHEREUM MATTERS 97

CHAPTER 10: UNDERSTANDING CRYPTOCURRENCIES ... 105

CHAPTER 11: ETHEREUM USE CASES 109

CHAPTER 12: WHO SHOULD INVEST IN ETHEREUM AND WHY? 118

CHAPTER 13: THE BLOCKCHAIN 146

CHAPTER 14: CONS OF CRYPTOCURRENCY 168

CONCLUSION 181

Introduction

Bitcoin was the first digital currency, created in 2008, and sparking a technological movement of the likes never seen before and likely never to be seen again. Ripples are still flowing across the financial industry today but there is more to come.

Bitcoin was the very first technology that allowed us to send money securely over the internet, without any worries about censorship or fraud. However, it didn't take long for a few pioneers to see the true potential of the blockchain, the technology that powered the payments system. What allowed us to send our payments securely also had the very real potential to be responsible for the decentralization of the internet.

But Bitcoin was never designed for sending any more than a couple of kilobytes of data in each transaction and neither could it perform any calculations that didn't fit the scripting language, which was somewhat limited. It was the belief of the Bitcoin creator, Satoshi Nakamoto, that by putting in a limit on the function of the currency system, security would be much better.

Enter Vitalik Buterin, the person responsible for Ethereum. He had a unique way of looking at things. He envisioned Ethereum as being a "world computer", a computer that would fit a virtual machine, a token called an Eth, a Turing-complete machine and "fuel" that powered each transaction into one network. This unique combination allows for the creation of smart contracts and DApps, or decentralized applications.

The Ethereum version of the Bitcoin is called the Ether or ETH token and, like the Bitcoin, it can be used for paying machines and people. Essentially you send ETH to a smart contract and this will then self-execute and carry out a set of complex instructions.

Across the course of this book, you will learn more about what Ethereum is, how it works and a practical example on drawing up a smart contract. You do need to have some background knowledge on computer programming for this.

Chapter 1: What Is Ethereum

Ethereum's blockchain is public, programmable, and even decentralized. Ethereum provides a peer to peer contract as it allows you to mine and trade ether, the platform's cryptocurrency token. You will be surprised to know that Ethereum was not proposed until the year 2013 by a Russian Canadian engineer that worked with bitcoin. Taking his inspiration from bitcoin, Vitalik Buterin used it to create Ethereum. His idea was to create a blockchain that would exceed bitcoin. However, it was not until the following year that Ethereum got the funding that it needed to begin the development process.

The official logo of the Ethereum platform

It was during the creation process that developers decided that Ethereum was going to go beyond bitcoin's peer to peer system and would have more services to offer its users than bitcoin. However, there were questions that came up about the scalability and the security that Ethereum had to offer. These are still two of the most prominent issues that the users of Ethereum have. As Ethereum was being developed, it won the World Technology Award in 2014.

The live blockchain was launched in July of 2015. At first, the program was developed by the Ethereum Switzerland GmbH and the Ethereum foundation. It was in spring of the next year that ether was established, and its net worth was one billion dollars. A website by the name of Vox said: "Ethereum is a new digital currency and is a challenge to bitcoin because of the wide range of services that bitcoin was unable to provide."

There are a lot of people that do not know that there are more cryptocurrencies than Ethereum and bitcoin. However, why do you want to choose Ethereum over any other cryptocurrency? Below you will see the good and the bad of Ethereum so that you can make an informed decision on if you want to invest with Ethereum now or sometime in the future.

Your contract will be executed as you state it to be

Many times, when a contract is written, a lawyer has to look at it, and a judge has to enforce it. This will be an expensive process. So, Ethereum offered smart contracts as a way to provide a cheaper contract solution. The smart contracts will be governed by the distributed autonomous organization.

Because of the distributed autonomous organization, you are not going to have to worry about how your contract is written or if it will be carried out as you want it to be. Each contract will be required to work inside of the DAO's rules. This way both parties involved in the contract are protected.

Because of the DAO, there is not going to be any need for judges or lawyers to be involved in your contract process. At the same time, you need to make sure you are rereading your contract before you are sending it to the system. A trustless operations system has never been established; but, Ethereum is using the way that technology evolves to attempt to create one.

Distributed Autonomous Organization

Digix was funded by five million in ether so that another cryptocurrency called dapp could be created. This cryptocurrency is backed by gold which makes it different than any other cryptocurrency. Dapp's funding was raised in one day. Thanks to the funding, it was able to start the company right away without having to wait for the financing. Once they started their business, they were able to create a board that determined how tokens would be distributed and various other things that would require someone to make the decision. This took investors, banks, and lawyers out of the equation. While for investors being left out of the equation is not fun, it does take some stress off of them because it is terrifying to invest in a company that they do not know what will happen with them.

The DAO will take the options that are required for the contract and simplify them into a single layer, so developers and users have an easier time in working on the platform.

Little costs

Being that Ethereum works off the DAO, it will eliminate costs that will be associated with business functions due to the fact that they will be done automatically. This is mostly due to the fact that you are not going to have to purchase an office building as you will see later on in this section.

EtherEx will provide you with a decentralized system that will have a cryptocurrency that is trustless. The EtherEx will work on an infrastructure that is similar to Digix where there will be a group of people that make important decisions. This board and the foundation have assisted in reducing the gas cost for a nonprofit organization.

What it comes down to is that when a business is built on a DAO, Ethereum will reduce the cost of setting up a building. You are no longer going to have to have your own building or buy office supplies for your employees to use. There will still be a small cost that will come with using the DAO. However, that cost is not going to be nearly as much if you had to rent your own building and all of the supplies. On top of it, you will have an unlimited number of employees working for you since they will be working remotely.

Still new

Sadly, Ethereum is still new, and it will continue to be developed so that their users can hopefully have a platform that can offer them anything that they could possibly want. When you compare Ethereum and bitcoin, you will see that bitcoin is more established, but it cannot provide what Ethereum can offer.

Legal advice

Ethereum has gotten rid of the need to involve judges and lawyers in your contracts. This can be good because you will be saving money. But, you have to know how the system works to ensure that you are making a decision that will be the best for you. The only thing you will have to worry about is that computers will have flaws, so a human has to be there to ensure that the platform and its code are running efficiently.

Changes

With the system always updating you will have to deal with the system shutting down so that it can apply those updates. The servers can also end up being overloaded which will cause them to shut down without warning. The evolving is not always a bad thing, but it will be if you are working on a deadline and you are not able to meet your deadline.

Remote employees

Since you will have employees that are not going to be working where you can keep an eye on them, you will have to deal with them being in different time zones and even with them being distracted by their personal life. It is not going to be too different than if your employees were working in an office building. So, make sure that you understand what is going on with your employees so that you can better prepare for what they will be dealing with as they work for you.

Competition

One of the biggest things that you will have to deal with is competition. There may not be enough competition, but if that happens, then you will be around for very long. But, when there are too many people on Ethereum then you will have a lot of people to fight against when it comes to mining pools.

There will be plenty of good things that will come from using Ethereum just like there will be bad things. But, you cannot stop that from using Ethereum. Ethereum is using technology and evolving to create an entirely new world of digital currency. You never know what will happen with Ethereum as it is changing and you will be part of history when you use Ethereum.

Take the good with the bad and figure out if Ethereum will be the platform you to invest in or not. If you can get past the bad and move on to the good, then you will be able to invest with Ethereum without any worry.

Chapter 2: Opportunities To Invest In Bitcoin (Classic And Cash) Vs. Ethereum

Direct investing

The first type of investing, and often the one that is seen as the easiest right now, is to work with direct investing. What this basically means is that you will take your American Dollars (or the other currencies that you want to use) and then use these to purchase some of the cryptocurrency of your choice. Investors hold onto the cryptocurrency, not spending it or using it for anything else.

After a bit of time, the investor is going to exchange the cryptocurrency back for their original form of currency, such as the American Dollar. The point is that the cryptocurrency is going to gain value and then you will have more money. So, if the cryptocurrency that you picked out was worth $10, you would hold onto it for a bit (sometimes for a few years to make the most) and then when you sell, it may be worth $20. You can exchange out the Ether or the Bitcoin that you have in order to get it back in USD and make a profit.

This is a pretty basic form of investing, especially if you are going to stick with the investment for the long-term rather than switching right away. Both of these platforms have seen a lot of growth ever since they started and since it is predicted that they will grow you can easily make a big profit. However, you need to stick with it for the long-run. Like any other investment, there will be highs and lows over the short-term so this is not always the best option if you want to make money right away, but if you can wait it out for a year or so, this is a fantastic way to really see a return on investment.

There are many reasons why buying cryptocurrency could be profitable. Despite the fact that this currency did have a rocky start because it was a new idea, there are many retailers around the world, like Overstock and Microsoft, who already directly accept Bitcoin, and there are even ways to work around the problem with some of the retailers who don't. For example, you could use your Bitcoin to purchase gift cards that work on Amazon.

It is believed that proponents of cryptocurrency are going to keep on growing as time goes on. Cryptocurrency is fast and easy to use and many major companies are already using it. And as cryptocurrencies grow into new markets and become more widely accepted, the value of them will grow as well.

This is a good thing for the investor. As more and more people start to accept the value of cryptocurrency and it starts to be used in more companies the value is going to keep growing. While it has already increased quite a bit since it was developed and introduced, there is still a lot of potential throughout the world. It doesn't really matter which country you are from or which currency you use, Bitcoin goes all around the world. This opens up a huge market and if you start investing early on, you will see your money grow.

There are still some people who are wary about direct investing in Bitcoin or Ethereum options. While the price of these is going up, there are some that are worried because it can be almost impossible to figure out what a fair price for the Ether or Bitcoin will be. Part of what will make assets, including currencies, valuable is that they already have a history of appreciation, which cryptocurrencies have not had the chance to work with. Then there is also a problem with people not agreeing on the right rules for Bitcoin, which is why there has been the separation of Bitcoin in recent times.

And defining these currencies is hard, which can mean it impossible to figure out their value. Since a currency is something that is backed by the government, which cryptocurrencies pride themselves on not being, and they are not really stocks, they aren't going to report their earnings or generate a profit, two things that help others figure out a fair price for the currency.

Now, there are a few people who are working on formulas that will help consumers figure out what the Bitcoin fair price would be. But these conflict quite a bit. One financial analyst from London believes that the value of bitcoin is over 200 percent higher than it should be but there are some skeptics who believe that there really isn't any value in this cryptocurrency to start with.

The biggest issue that still surrounds these cryptocurrencies is that they are associated often with criminal activity. Since these currencies are all anonymous, it is easy for people to purchase items that are illegal or do things that are illegal without anyone being able to figure out who they are. This has put a bit of a shadow on the cryptocurrencies, but for the most part, people still like to use the currency and most people use it without having any criminal intentions.

Bitcoin and Ether are great options to choose to invest in, you just need to understand what you are getting into. One of the biggest downfalls is that the currency is still new and many companies are not accepting it and that they do have kind of a criminal history behind them. But there is a rising popularity of this type of currency, which could mean a big profit for you. Just make sure that you watch the market and do your research instead of just jumping right in.

Lending money to others

Another option that you can choose to do with the Bitcoin and Ether is to become a lender. This is discussed a bit later in terms of lending out to startup companies through the ETH platform, but this method would be more to individuals. You can help them with a small business idea, but it would be more about helping with a small car loan or helping them with some bills and so on.

There are a lot of people who may need a little help with their bills or other things in life, but for one reason or another, they are having trouble doing it through a bank. If you have a few Bitcoin available, you may be able to make a bit of money and help someone out.

Before you lend the money, make sure to take a look at the background of the person you want to lend to. What are they going to use the money for, how much do they want to borrow, how long do they want to pay it back, and what are some of the reasons that the bank didn't give them the loan? If they are delinquent on five loans in the last few years, they probably aren't the best option, but if their credit score is a little low because they are young and haven't built it up, they may be fine.

Make sure that you and the borrower set up the rules of the transaction before getting started. You should set up how long they have to pay it off, how much they pay off each month, the interest rate, and what happens if they don't pay it back. Working with a smart contract can be a great idea because it ensures that both parties will be protected.

Indirect investing

There are a few different types of investments that you can do that will help you make money off Bitcoin and Ethereum, but which won't require you to actually purchase the currency and hold onto it. Some of the options include funding a startup for an application that is on the Ethereum platform, investing in the Bitcoin and Ethereum companies, using smart contracts and more. Let's take a look at how these all work so you can make an educated decisions for your investing.

Investing in the company

The first type of indirect investing opportunity that you can choose is to invest directly in the company. You will still not own any of the cryptocurrency with this, instead, you will invest in a way that helps the company to grow and do well. This is similar to what you would do when you join the stock market; you will pick out a company and then purchase bonds, stocks or other options that give you partial ownership of the company.

You will be able to pick how much you would like to invest into the company. This is generally considered a safe way to invest because these cryptocurrencies are growing like crazy and you have an actual tangible product that you are able to purchase and sell. You can choose what type of ownership you would like to have how much you want to spend, and more. Then every quarter, you will receive a profit when the company realizes profits based on the amount that you invested.

As long as the company is doing well, you are going to make a profit from this form of investing. You do need to watch the market to find out when the stocks are going up and when they are going down and plan your next move, but this is a really good long-term investment and it can become kind of passive. Since both

Bitcoin and Ethereum are doing so well and they have such a large market that is still untapped, it is not likely they will see a decline anytime soon. You could easily put your money into the companies directly and see the money come back over the years.

Using Ethereum to fund a start-up

Ethereum is quickly becoming one of the best places for tech start-ups to go when they need new funding. While this may have been done in Silicon Valley in the past, it is much easier to find this funding through the Ethereum platform. Many start-ups like this option because they are able to work on a platform that is distributed through the Internet, decentralized, and secure. And it has all of the ingredients that are needed so that a startup can get going without having to worry about their physical location.

To use this platform, users will have to use a liquid medium for value exchange, which is the Ether currency. They can also work with smart contracts, which is the common system for business and application logic through this system. To add to all of this, the Ethereum platform is great because it has developed a culture that is going to support and encourage innovation in the tech startup world.

Many new startups have moved over to this platform and have been able to sell their own 'tokens' to a globally distributed crowd of early adopters. These tokens are unique because they are cryptocurrencies that will have a special utility within a given application. In this case, the cryptocurrency is going to be used to help a startup get the funding they need to start and grow their business. For a project that is based on the Ethereum project, selling these tokens will solve two or three problems including:

It will provide the funding that the new company needs for their project.

It is going to bring together a new community of people who are excited, and who have an incentive, to get the application up and running.

It is a model of payment that can be used without having to worry about government or bank control.

This can be beneficial to the startup company and to the investor. The company is going to have a new way to start up their own business and get the funding that they need, without having to worry about going through the bank or another company to get the funding. This method is often easier and it can be nice to have a group of enthused people to help the project gain traction from the beginning, rather than just going through the bank.

The investor will also enjoy this as well. They will be able to find an application or a company that you are excited about and then can invest in them. You will be able to earn a profit when the company goes live or when the project is done, which will earn you money. And there won't be the constraints of banks and other investment options; you just require to work through the ETH platform and you will be set to go.

You do need to still do your research, though. There are quite a few startup companies that are using the Ethereum platform to help them grow their business or to start out a new project so there are many options for you to try out. But you do need to be careful. There will be some that are really good and some that are not that good.

While some of these options are going to be brand new to the world and may not have a bunch of research to show how they have done in the past, this doesn't mean that you can't do some research. It also doesn't mean that you should bypass these options. For these, make sure to get the information that you can. Take a look at the product, look at who runs the company, look at the business plan, and get the information that you can before you choose or give up on the investment.

This can help you to determine whether the company has a good future or not.

Start your own business

It is possible to use these platforms to start your own business. It would work similarly to how you would start a brick and mortar business, but you could keep it all online. There are really quite a few businesses who work on these platforms and the list is always growing. As long as you are able to provide services to others after accepting their cryptocurrency, you will be fine.

Some people decide to do a business completely online and then accept Bitcoin, Ethereum, and other cryptocurrencies as a form of payment. They can sell clothes, shoes, household appliances, and more. Then when they are all done, they can withdraw the currency into USD or another form of currency and use it how they would like. It is a pretty simple process and more companies are starting to add these on.

Even companies that don't offer services online are considering using Bitcoin as a form of payment. Grocery stores can accept Bitcoin by setting up a scanner that will take the money right from their customer's wallets. Some cleaning services, for example, could take these currencies online before the services. There are many ways that these cryptocurrencies can be used in helping you to start a new business or run an existing business.

The use of smart contracts

No matter which option you choose to do with investing, it is a good idea to learn more about smart contracts. These are useful because they are going to help you to exchange shares, property, money, and anything else that has value in a way that avoids having a middle man and avoids conflict. In most cases, you would use a lawyer to do this and then pay them and a notary to get the contract done. With the smart contracts, you are able to drop some cryptocurrency into the ledger, and the smart contract will be done.

Considering a real life case it could look like this: you could rent an apartment using the blockchain that you paid for with the cryptocurrency. You are going to get a receipt when the transaction is done, which will be held in the virtual contract. The apartment owner needs to send you a digital entry key by a certain date, and if this doesn't come to you, the blockchain is going to release your money back to you. If the key is released, then the money will be sent to the other party. This helps both sides to be safe. If the key isn't released, you will get your money back, but if it is, the apartment owner is sure to get paid as well.

There are several ways that you can use these smart contracts. The first way is to use it when you make sales and purchases with other people through these sites. Since all these transactions are going to happen online, it is best to have some sort of safety feature to protect both parties.

You are sure to get paid no matter what you do, and this keeps you safe.

If you have some computer knowledge, you can set up a system to create these smart contracts. The users of these contracts have to pay a few Bitcoin to make them work, so you could make a bit of money off each these transactions. You would need to be versed in coding to make the contracts and make sure that they are binding and working well, but it can be a way to make a second income.

Mining

One popular option for those who want to make money with these currencies and who have some programming knowledge is the process of mining. In most cryptocurrencies (especially Bitcoin), there is a set amount of the currency available and there won't be any more that is ever made, unlike what we can find with most traditional currencies where the government can just make more. But when these currencies were released, they only released a certain amount. The rest need to be mined to be released.

Those who are able to do the mining business are going to be able to help not only themselves in this process but also other users. Remember that Bitcoin and Ethereum are used with a blockchain to help keep information safe and secure, but since it posts on a ledger, it needs to be done in a way that shows that the transactions have been completed. Both of these have a complex system for how these transactions should be stored, without giving away any information about either party.

For example, Bitcoin will have a series of numbers that will be related to each other. This means that if you change one number in the sequence, all the numbers that follow will change as well. There are many other rules to it as well, which is what will make mining so hard. If you are able to successfully complete some of these sequences though, you can earn money in

Bitcoin so for programmers and computer buffs; this is the best option to use.

One of the big advantages of going with the indirect investments is that there are so many options that you can choose from. You can really personalize the option and figure out what you like. You can choose to work on the contracts and on mining if you would like to stick with more of the programming stuff, you can start your own business, you can invest in a start up or do so many other things. The options are what makes these platforms so great to use.

The disadvantage is that you do need to do your research. These options are sometimes hard, like with mining, and you are not going to make money overnight. And with some of them, like with investing in a startup company, you may lose money, even though the Etereum platform is growing. You have to always be vigilant in what you pick out and make sure to protect your investment as much as possible.

Mixed investments

Doing a mixed investment is probably one of the best decisions that you can do with cryptocurrency investing. This allows you to split your money up between a few options in the direct investing and the indirect investing categories. This reduces the risk, helps you to get the benefits of each, and can bring in more money over time.

The idea behind this one is to not put all your money in one basket. Maybe you take some of your money and split it between investing in the Ethereum platform and the Bitcoin platform. Then you take some of your money and fund a startup and a few other companies. This helps you to spread out the money and keeps you safer than before.

If something does happen to one of your investment options, such as picking a startup company that doesn't do as well as expected, you still have some other investments pulling you up. You are likely to still make a profit, even though you lost a bit of money. But if you placed all the money in that investment that tanked, you would have lost out on everything.

There is no right or wrong mix of your investments to see the most money, but you should take a look around and pick out the one that you are the most comfortable with. This process is often known as diversifying your portfolio and it is one of the best techniques to make sure that you reduce your risk. All the best investors, whether they are working on cryptocurrencies or something else, will use this option because it helps to minimize their risks while maximizing their profits.

Beyond 2017, anything can come about. Indeed, even industry specialists can't precisely anticipate how exceptionally quick cryptocurrency expansion will wipe out. Plenty will rely upon whether autonomous countries take the Japanese route and distinguish them from genuine currencies or the German course and transparently show hawkish intents.

We would all be able to concur that the innovation is profitable and here to stay. The million-dollar question encompasses adoption. Will controllers around the globe grasp all the blockchain currencies as fiat measures up to, or will they regard them as antagonistic foes to their monopoly-based fiat frameworks?

How to purchase & store cryptocurrency?

On the off chance that you have some play cash and need to make a gamble on cryptocurrency, you ought to completely feel 100% good with using up all that cash. Cryptocurrencies have collapsed some time recently, over and over again, and presumably will again later on. They're likewise generally costly in the event that you should get a few, you may be served by sitting tight a bit at costs to drop, so will probably get it. There are hordes of modes to purchase cryptocurrencies, and a few nations have even set up methods to buy them by means of an ATM.

Coinbase is amongst the extremely eminent Bitcoin agents and frequently prescribed for beginners. Coinbase permits you to purchase Bitcoin and different cryptocurrencies by connecting to your debit or credit card. Business Insider testifies that the mobile app is buggy, and banks will occasionally bolt a card subsequent to making these transactions. With that in mind, BI suggests telling your financial institution prior attempting to make a buy.

There are a couple of different choices; however, they have to a lesser extent a reputation: Kraken is one respectable option; it has been around since 2011 and works with an extensive variety of dealers & governments. There's likewise Gemini, yet it is not thus far accessible in each state.

Lastly, since trades, even the biggest ones, have smashed unexpectedly, it's additionally imperative to get yourself a protected place to store your Bitcoin, in the event that your supplier leaves the business or endures a hack. These tools are frequently alluded to as Bitcoin or cryptocurrency wallets. The wallet is, in fact, a physical gadget that interfaces with your PC and goes about as another source of defense. This implies you can't send Bitcoins from your wallet without possessing the physical gadget. There are so many of them such as:

Hardware ledger wallets

Ledger Bitcoin wallet

Coinbase Bitcoin wallet

TREZOR Bitcoin wallet

Blockchain.info Bitcoin wallet

Jaxx Bitcoin & Altcoin wallet

Exodus Blockchain assets wallet

MyCelium Bitcoin wallet

Bitcoin Core wallet

Keepkey Bitcoin wallet

Electrum Bitcoin wallet

Xapo Bitcoin wallet

Armory Bitcoin wallet

CoinKite Bitcoin wallet

Bitcoin Wallet

Green Address Bitcoin wallet

Bitcoin Wallet

BitGo Bitcoin wallet

Copay Bitcoin wallet

Airbitz Bitcoin wallet

CoolWallet Bitcoin wallet

BitLox

Software based online ledgers

SAP ERP

Sage 100c/300c/X3

MYOB

InFocus

Intacct

Acumatica

NetSuite ERP

Oracle Financials Cloud

FinancialForce

Goldenseal Accounting

Unit4 Business World

Traverse Cloud ERP

Microsoft Dynamics

In case you're earnest regarding crypto, obtain a hardware wallet. It doesn't become so clearer than this caption. Of course, hardware wallets cost cash and nobody loves to pay out money on items they can acquire for free. On the contrary, the measure of security you obtain by utilizing a hardware wallet is substantially more significant than the 50-$100 you'll pay for buying the genuine gadget. Don't say "it won't transpire" in light of the fact that once cryptocurrency progresses toward becoming standard (and it appears as though we're arriving really quickly) there will be numerous more instances of hacking or burglary. Ensure you're prepared...

What About Alternatives?

One could contend it is similarly conceivable to put resources into vehicles, for example, GBTC. That investment alternative is additionally straightly connected to the Bitcoin esteem. That is unquestionably a substantial alternative, supposing you are an accomplished investor or financial investor. The normal individual in the city won't consider GBTC to be an essential alternative to any methods. Additionally, GBTC likewise gives access to Ethereum Classic, which isn't the equivalent as Ethereum. This may puzzle many people who aren't knowledgeable in the fine complexities of cryptocurrency right now.

Standard customers hoping to receive the benefits from Bitcoin's value additions should investigate BitcoinIRA. It is an amazing answer for individuals who need to procure cash as an afterthought without worrying. Furthermore, it is an extraordinary tool to develop a retirement fund. The Bitcoin cost will in all probability just go up more from here. BitcoinIRA is an available as well as a reasonable solution with a capability for high returns.

During 1999 the world perceived the IPO (speculative internet). These days, it's the ICO (initial coin offering). Organizations based on the blockchain, a digital database for recording monetary exchanges and different sorts of agreements, are boosting money by means of offering digital "tokens" that can characteristically be utilized to pay for goods & services on their platform or merely put away as an investment. Up to this point in 2017, organizations have brought $180 million up in ICOs, contrasted with $101 million all of a year ago, as per Smith + Crown, a blockchain research, data and consulting group. Time and again, these are early tasks that are a long way from producing substantial income.

Enthusiasm for cryptocurrencies is achieving the majority. In the meantime, 10 monetary establishments joined with cryptocurrency platform Ripple just recently to send real-time international payments, joining a list of customers that by now included Bank of America & RBC. A definitive visualization is the world in which all data & transactions are detectable by means of an electronic ledger that wipes out hold-ups caused by divergent monetary standards and money related frameworks. Blockchain as of now claims to process 160,000 transactions per day crosswise over 140 nations.

Despite the fact that the vast majority of the folks purchasing Ether & Bitcoin are a singular investor, the advantages that both have encountered have taken what was in recent times a peculiar periphery try into the domain of huge cash. The consolidated estimation of all Ether & Bitcoin is currently worth greater than the market estimation of PayPal and is moving toward the span of Goldman Sachs.

Various Other Significant Cryptocurrencies for Investing

Litecoin (LTC)

Litecoin commenced in the year 2011, was amongst the underlying cryptocurrencies following Bitcoin and was regularly alluded to as 'silver to Bitcoin's gold.' Litecoin depends on an open source global payment network that is not controlled by means of any central authority and utilizes "scrypt" as an evidence of work, which can be decoded with the assistance of CPUs of consumer grade. Despite the fact that Litecoin resembles Bitcoin from numerous points of view, it has a speedy block generation ratio and consequently offers a speedier transaction affirmation. Other than developers, there are the greater numbers of merchants who acknowledge LTC.

Digital Cash (DASH)

Dash initially identified as Darkcoin is a more cryptic form of Bitcoin. Dash imparts greater obscurity as it attends to a decentralized mastercode network that formulates exchanges untraceable. Released to market in January 2014, Dash had a notable success and gained lots of supporters in a very short time. In March 2015, "Darkcoin" was rebreeding to Dash, which appears for Digital Cash and works under the ticker DASH. The rebuffing didn't transform any of its technological characteristics, for example, InstantX, Darksend.

Zcash (ZEC)

Zcash is decentralized & open-source cryptocurrency initiated in the last part of 2016, seems promising. On the off chance that Bitcoin resembles http for cash, Zcash is https, is the manner by which Zcash characterizes itself. Zcash offers privacy along with selective transparency of exchanges. Along these lines, similar to https, Zcash asserts to give additional security or protection where entire transactions are recorded and distributed on a blockchain, however particulars, for example, the sender, beneficiary, and sum stay private. ZEC offers its clients the option of shielded transactions, which take into account content to be encoded utilizing advanced cryptographic strategy or zero-knowledge proof construction known as a zk-SNARK created by its group.

Monero (XMR)

Monero is safe, private & untraceable money. This open source cryptocurrency was propelled in April 2014 and shortly spiked incredible enthusiasm amongst the cryptography community and fans. The advancement of this digital money is totally donation-based & community-driven. XMR has been started with a solid core on decentralization plus expandability, as well as empowers complete privacy by utilizing an exceptional procedure described as ring signatures. With this strategy, there shows up a gathering of cryptographic signatures consisting of as a minimum one genuine member yet since they all seem legitimate, the genuine one can't be disengaged.

Ripple (XRP)

Ripple is a real-time worldwide settlement network that tenders instant, definite and economical international payments. Ripple empowers banks to settle trans-border payments in real time, together with end-to-end transparency, and at reduced prices. Circulated in 2012, Ripple money has a market capitalization of $1.26 billion. Ripple's consensus ledger its strategy for adaptation needn't bother with mining, an attribute that diverges from Bitcoin as well as Altcoins. Seeing as Ripple's structure doesn't necessitate mining, it diminishes the utilization of computing power and limits network latency. Ripple trusts that distributing value is a commanding approach to boost certain practices and in this way as of now intends to disseminate XRP fundamentally in the course of business advancement bargains, inducement to liquidity suppliers who present extra tougher spreads for payments and selling XRP to institutional purchasers keen on putting resources into XRP.

Chapter 3: The Double-Edged Sword Of Ethereum

If Ethereum was objectively better in every way in comparison to Bitcoin, we would have seen the evolutionary extinction of Bitcoin and the total dominance of Ethereum relatively quickly. Since that is not the case and since both of them are currently in existence simultaneously and in good number, we must evaluate the pros and cons to each. Even though both are built on the same fundamental principle, the blockchain, they have different programs, and each come with benefits and detriments. Let's go over why you would like Ethereum, and then we'll tackle the negatives.

Contracts and the Interface of Ethereum

You can think of contracts in Ethereum as programmable accounts, but also in a legal sense as you would any general contract. They are essentially accounts you can interact with **via** Ether transactions. While contracts in the real world can require attorneys or even judicial figures to be implemented, in the crypto-world this is not the case. This allows for an extremely low cost and barrier for entry into the transaction world. This is due to the Distributed Autonomous Organization (DAO) as the governing body.

Every time you sign your name on a receipt or click the "I understand" checkbox upon purchasing a service, these examples, are the equivalent contracts you will see in Ethereum. However, these contracts in Ethereum can be made by anyone, and anyone can interact with them in an open sandbox like manner (at least in comparison to current transaction 'contracts').

More thoroughly, a smart contract is basically a computer program that has been written in an Ethereum high-level programming language. These languages are primarily Solidity or Serpent. The program is then deployed into a special transaction along with a minimal transaction fee. When this transaction is executed and deployed, the program is then securely stored in the blockchain and can exist forever.

These types of contracts can enforce or mediate, in a sense, any deal taken between two or more parties. As an example, the recently advanced development of the Internet of Things, we can now design a microprocessor in our car that has the purpose of mediating and enforcing smart contracts. By having this processor installed in your vehicle, if you intended to sell it, but a specific buyer is unable to pay until a pre-specified (or unspecified!) later date, you can write up your own smart contract such that when a buyer deposits enough Ether into your wallet, the private key will be unlocked for use and he can drive the car. If the contract is not fulfilled, then the engine will only start when you personally use your working, private key. In the future, contracts can be enforced automatically and specifically by perfectly logical computers and algorithms.

In our current legal system, most contracts are often written in ambiguous, sometimes arbitrary language and words which are subject to our own interpretations. If you and the opposing party differ on interpretations of a certain contract, the process becomes painful and will likely require court cases which will obviously further be evaluated on a subjective basis. Juries and judges are not omnipotent and can make erroneous judgments. Here, with smart contracts, the contract is written in a logical, nearly mathematical way in which interpretation is no longer a problem. With smart contracts, which will be written in a clear and universal programming language, the question of who is correct and who is incorrect is no longer an ordeal as the contract leaves no room for interpretation. Once the contract is deployed into the blockchain, you can trust that the contract is protected and the execution of the contract will take place in a logical, rational, and **mathematical** manner.

Gold, Crowd Sales, And DAOs

As discussed before, people feared when the dollar became unhinged from gold, as they thought it would likely collapse the actual value of the dollar. Of course, this was not the case, as the value of the currency is a psychological, collective agreement only. However, with commodity currency that is backed by a physical asset such as gold, the confidence in value that the consumer or holder has become stronger.

Let's look at an example to illustrate my point. If half the world's population were to move to Mars and start their own currency, everyone would have to agree on how much a single token is worth. This makes very little sense to do until you make it backed by a commodity. Deciding the value of a Mars Token or Moken **requires** you to give it value to a commodity, like gold, water, or oil. You could begin distributing your currency after everyone agreed that a Moken represented one liter of water, or oil, or a kilogram of gold. Over time, the value of a Moken would fluctuate until a global Mars consensus had been reached (generally speaking). However, to actually be a commodity currency, there must exist a central bank that holds enough water (or oil or gold) to exchange every Moken currently on the market for its respective commodity value. When this is the case, the confidence in the value of currency increases, and people worry less about inflation and other monetary issues.

Thus, although it is not required, the initiative to produce a gold-backed cryptocurrency makes sense to promote consumer and user confidence in a currency's value. Using a **contract** through the DAO, a crowd-funded project termed Digix was produced which backs all of the registered Ethereum coins with a value of gold. To show the power of Ethereum's community and crowd-funding efforts, a total that exceeded $5 million USD was funneled to Digix in a relatively short amount of time – completely funded with Ether.

DAOs, Dapps, and Crowd Sales

The use of the DAO crowd sale presented a new distributed app (Dapp) that was quickly and efficiently funded to the total of the $5 million project. The project was completely set up by the board of management and required no legal attorneys or lawyers. If (or when) cryptocurrencies become universally adopted as a major form of currency, the ease of producing large funded projects will become much quicker since the funds are not allocated to any region, the laws and regulations that were once prohibitory for many are now sparse and easily maneuvered around. Therefore, developing ideas in the realm of Ethereum, or other cryptocurrencies, will streamline funding and financing for any project generated.

Ethereum itself is based on a DAO and can theoretically reach zero cost to use in all scenarios because functions are done autonomously via the contracts. In reality, there will always be costs associated with any execution of the smart contracts, so there is always a small drain on all activities done within Ethereum. The point of Ethereum, however, is to minimize these costs considerably in comparison to physical labor, offices, and the infrastructure to run full businesses. Thus, executing a contract in the space of Ethereum is going to be significantly lower than executing the same contract through traditional means.

The Real Benefits of Ethereum

Imagine instead of having employees, housed in skyscrapers, with expendable resources, having to go through massive amounts of transactions and "contracts" done between businesses at their cubicles versus a dispersed and distributed peer-to-peer group that largely operates through technology alone and all "nodes" of transactions simply take place on the contracts within a specific business. The costs then become astronomically smaller to execute business and financial functions that most companies have entire departments allocated for today.

The cost in Ethereum is called "gas, " and it is a universally accepted cost for doing any computational work by the Ethereum base. Gas has a constant cost, regardless of the volatility of Ether's value, helping to stabilize the currency and enforce more efficient paths for coding the contracts.

In essence, the real purpose of Ethereum is to create a platform in which methodologies can be built upon to run business transactions and costs much more efficiently than that in use today. For instance, the minimum credit card purchases you can have anywhere (if you're lucky) is 99 cents. That cost is directly attributed to the cost of allocating your account's balance to another account's balance. There is always a net loss of money from any transactions from the consumer to the producer in the case of digital currency transactions because some of that cost is required to go to the corporation that is in charge of moving that money (Visa, Master Card, American Express, etc.). This cost becomes prohibitive for micro transactions and also accumulates linearly with some transactions processed. You can think of normal currencies as bulky, weights that can't be divided much lower than 99 cents, or the transaction **itself** becomes a net expense for someone within the transaction.

With Ethereum and other cryptocurrencies, you can imagine that the tokens themselves are infinitely divisible and that the transaction costs can and will become negligible in comparison to that of today's standards. Currently it is impossible to be able to pay-per-second for any type of streaming entertainment because if you were to open a transaction or contract to watch only a second of video and subsequently close the transaction, requiring the payment to be executed, and if you were to do this continuously, hundreds or thousands of times, this would cost the video streaming service more money than they would be producing from you because of the transaction costs themselves! This type of issue is averted with cryptocurrencies, as the infinite division is not an issue, and transaction costs become minimized.

With respect to Bitcoin, Ethereum is more intended for the infrastructure of businesses and intelligently designed contracts to execute business needs and financial transactions. It can be thought of as an alternative to the skyscraper filled with cubicles. Its infrastructure was specifically designed with this in mind, as a "black box" in which to operate transactions through. Like Bitcoin, there is little stopping Ethereum from being an alternative currency to fiat and commodity currencies. You can conceivably trade anything using Ethereum, but this is not Ethereum's strength in comparison to other cryptocurrencies (CCs) – they can all do this. It's rather the computing language that allows the smart contracts to exist that makes Ethereum more valuable than BTC (in my opinion).

The smart contracts that are stored within the blockchain of the Ethereum platform are protected by the platform system similar to how a transaction within the Bitcoin blockchain is executed and subsequently protected by the computational power of the entire network. Since Ethereum is decentralized and all of the transactions are verified at each of the computers in the P2P network, there is no single third-party that has enough power to destroy or modify an existing contract.

In essence, the fundamental benefits of investing into Ethereum is the cryptographic nature, it's anonymity, it's universality, it's divisibility, and most specifically, it's coding language which specifically targets it for automated transactions and contracts.

The Drawbacks

We have gone over the positives of Ethereum, and like any investor, you must also know the risks involved. Understanding the entire paradigm of Ethereum requires you to understand its weak points in addition to its strengths. Ethereum's most obvious weakness is its age. It is a very immature cryptocurrency, and major portions of the infrastructure are still being developed. While Bitcoin has a large market cap, this is largely due to it being the first cryptocurrency and Ethereum is growing quickly.

While Ethereum is meant to drop the need for attorneys, lawyers, and judges from the expenses required to produce contracts and execute transactions, people are still going to be a part of the process. Ethereum itself is built upon human intention and innovation and thus will always rely (and be threatened by) on the human mind and input. If there are major flaws or updates required for the servers, the entire market can be taken down or act erroneously, which would weigh heavily on people's perspective of its resiliency and thus depreciate its market value. This is not unique to Ethereum, but it is probably due to its immature stage of development in comparison to Bitcoin.

If for instance, major banks began to develop contracts within the Ethereum blockchain platform and a major flaw in the design was identified, the network may be taken down, the bank(s) would likely withdraw their funds, and the market cap and market sentiment would depreciate significantly, thus lowering your investment's return. Thus this stage of Ethereum is likely to be volatile, something that can scare the emotional investor, but has an exponential capability to produce massive gains for ROI.

Although not unique to Ethereum, building business models around Ethereum with the goal in mind of minimizing infrastructure costs of your company by removing the office space and cubicle setting can impact workforce efficiency. It is not currently a large trend for bankers to allow the work-from-home paradigm, but it's not impossible for this to occur on a large scale. This will be in the hands of large corporations to weigh the risks and rewards of decentralizing their corporation's contracts and transactions against keeping the business centralized and homogenous in the organization.

As discussed above, to make informed decisions on investing, it is required for you to know both risks and rewards. By focusing solely on the possible rewards, you are very likely to make massive mistakes with your money. Instead, identify the probability of success in every scenario you choose to put an investment in. With Ethereum your investment thesis should be somewhere along the lines of

A belief that CCs are inherently valuable

That more people, corporations, and governments will adopt the use of CCs

That evolutionary-style progress will continue to advance the uses and applications of CCs

That Ethereum presents a benefit to business models by reducing infrastructure and overhead costs

That Ethereum has enough security and resilience that it is comparable or better than Bitcoin

That short-term volatility of market prices is manageable as long as the above five thesis still ring true

Chapter 4: What Is Ethereum?

Ethereum is an open and decentralized blockchain platform that allows its users to run decentralized applications on blockchain technology. Similar to Bitcoin, no one owns or manipulates the Ethereum protocol. It is considered as an open-source project that allows the use of smart contracts.

Ethereum was developed and proposed by Vitalik Buterin back in 2013. It received its funding from a crowdsale in 2014. A year after or in 2015, Ethereum was finally launched in the market.

Ethereum has long been considered as the second most successful cryptocurrency in the world next to Bitcoin. As of January 14, 2008, the price of 1 ether (Ethereum token) is around USD 1,430.

What is a **Cryptocurrency**?

Before we continue to discuss the details behind this cryptocurrency, you should first understand what a **cryptocurrency** is.

A cryptocurrency is a type of digital asset. It is held and stored electronically (online). Take note that it does not have a physical existence. A cryptocurrency differs from fiat money in the sense that fiat money is the recognized and official currency of a state like the US dollar. A cryptocurrency is also not considered to be legal tender. In law, legal tender refers to that which the debtor can compel his creditor to accept payment. Of course, the exception to this rule is if it is stipulated in the contract that payment may be made in cryptocurrency.

Cryptocurrency is secured using **cryptography**. This refers to the practice of turning information into codes. Cryptography was extensively used during the Second World War since it was important to ensure that the security of information communicated in the army was protected against enemy spies. This is how trustworthy and reliable cryptography is.

Today, there are more and more individuals and merchants who accept payments in cryptocurrencies, such as Microsoft, Overstock, Steem, Fiverr, Virgin Galactic, and many others.

Who is Vitalik Buterin?

Unlike the founder of Bitcoin whose identity remains anonymous, the co-founder of Ethereum is well known to the public. Vitalik Buterin is said to be the mind behind the Ethereum cryptocurrency. He is a young Russian-Canadian computer programmer born in 1994. He is also a co-founder of the famous, Bitcoin Magazine. It was his father who introduced to him the world of cryptocurrency. His father taught him about bitcoin when Vitalik was only 17 years old. Since then Vitalik placed his focus on learning more about the technicalities and codings behind cryptocurrencies, which soon led him to come up with the white paper for Ethereum.

What are Smart Contracts?

Smart contracts are a form of computer protocol that can execute contracts provided certain conditions are met. As such, they effectively enforce the execution of contracts and ensure that they are executed according to the prescribed conditions. Smart contracts cannot yet handle complicated tasks, but you can always use many smart contracts to handle and fully execute such tasks.

The term **smart contracts** is nothing new. It was first coined back in 1996 by Nick Szabo. The concept was developed over the course of several years. Back then, Nick Szabo defined smart contracts as follows: New institutions, and new ways to formalize the relationships that make up these institutions, are now made possible by the digital revolution. I call these new contracts "smart," because they are far more functional than their inanimate paper-based ancestors. No use of artificial intelligence is implied. A smart contract is a set of promises, specified in digital form, including protocols within which the parties perform on these promises.

Ethereum as an Altcoin

Ethereum is considered as an **altcoin**. An altcoin is simply a term that is short for **alternative coin**. In the cryptocurrency market, all cryptocurrencies that are not a bitcoin are considered as altcoins. To date, there are more than 1,000 altcoins that have already been developed and launched in the market, and the number keeps on increasing. Ethereum is one of these altcoins. However, unlike the majority of altcoins in the market, Ethereum, is one of the most successful, popular, and profitable cryptocurrencies in the market today. In fact, many experts claim that Ethereum will soon be able to take Bitcoin's place as the number one cryptocurrency in the world.

It is worth noting that there is nothing wrong with being considered as an altcoin. In fact, many investors these days are more interested in altcoins than in bitcoin.

Altcoins like Ethereum usually have a big room for development; hence, their profit potential is also high.

Crypto Kitties

When you read about Ethereum, you will most likely encounter the term **Crypto Kitties**. Okay, these Crypto Kitties are a new invention launched in the latter part of 2017. They are like the classic toy, Tamagotchi, where you take care of a virtual pet. In this case, you will take care of a virtual cat that runs on the Ethereum Blockchain. This is an Ethereum-based game where you can take care of and breed a virtual cat. So, what makes this significant? Well, some people buy and sell these virtual cats. Just to give you an idea of how significant this matter is: People all over the world have already spent more than 6.5 million dollars on these crypto kitties. The thing is that you can sell them also at a profit. Based on the record, the **genesis cat**, which is said to be the most expensive crypto kitty was sold for $115,000.

If you look closely, the virtual cats themselves are just part of what makes this whole Crypto Kitty thing of any significance. The important thing to note is the possibility that it creates in the world of Ethereum and cryptocurrencies. This is just the beginning. In fact, this only signifies the possible uses and extent of Ethereum's blockchain technology. Imagine being able to sell and transact other things like stocks and gold, etc. using Ethereum in the near future. Well, it is worth mentioning that owning a crypto kitty can also be fun. It is like owning a Tamagotchi. However, this time, you can sell it at a high profit.

Chapter 5: Technology Behind Ethereum

In addition to the main ethereum blockchain protocol, there are also supporting technologies in development that seek to help the network, and components built on the network, run more efficiently. For example, whole new protocols are being constructed that aim to increase the functionality of distributed applications, while tools are evolving to allow these programs to harness data from multiple blockchains.

While there may be little that unites the following concepts on the surface, all are aimed at making ethereum more flexible for developers and users.

Whisper A communications protocol and tool set that allows applications built on the ethereum protocol to talk to each other; Whisper combines aspects of a distributed hash table and a point-to-point communications system. Whisper is best explained in practice as it can be used to help facilitate exchange by recording buy or sell offers, allow for the creation of general chat room-like apps or even provide "dark" communications between parties who don't know anything about each other.

With Whisper, you can imagine an ethereum application for whistleblowers who want to communicate to a journalist where they've stored a trove of data, but don't want their identity to be linked to that data.

Swarm

Swarm is a peer-to-peer file sharing system designed to efficiently store and retrieve data needed for use in ethereum applications and contracts. The easiest analogy to draw would be that Swarm is essentially BitTorren for ethereum.

As we will discuss later, storing data directly on the ethereum blockchain is expensive. While contract code will have to be stored on the chain, reference data needed for contract execution should not. For instance, if a simple contract were to say deliver an e-card with pictures, the photos would take up a lot of space. Perhaps a school would want to send out an album with photos of its latest graduating class. Such an application, if run on ethereum, might require a contract that is 1 KB, but is designed to deliver 1 GB of data. Storing and transacting that 1 KB of code might cost users a few cents, whereas storing the album itself could cost more. By instead storing the album remotely, and accessing the file via a BitTorrent-like system, this would allow ethereum applications to deliver the instructions, with the files to be transferred via Swarm, not the ethereum blockchain directly.

For smart contracts to execute properly, they need not just be a well-designed series of "if then" statements – they also need to know how to ascertain the accuracy of given inputs to those "if-then statements".

Mist

If ethereum is to be the new TCP/IP, the project needed a new version of 'browser', a usable frontend technology with which users explore the applications and offerings that utilize ethereum. Styled as a decentralized application discovery tool, Mist is meant to serve as a wallet for smart contracts that features a graphical interface and allows users to dynamically set transaction fees and manage custom tokens. Presently, Mist is still in beta and is under heavy development.

Chapter 6: What Is Ethereum?

Ethereum is a rising star in the world of cryptocurrency and is currently challenging the popular Bitcoin for the top spot. As of writing, it is second only to Bitcoin when it comes to value among existing digital payment methods.

After Ethereum's launch in 2015, Ether, which is Ethereum's currency, has exponentially risen in value. After only a year, ether's value increased to above 2300 percent. Currently, an ether (a unit of Ethereum) is valued at around 300 US dollars.

Ethereum Explained

Ether is second only to Bitcoin when it comes to value as a digital money format. It runs on a technology called Ethereum. This technology was first described by Vitalik Buterin in 2013. Buterin was a Bitcoin programmer and was only 19 years old at that time.

Buterin was once quoted as saying that he envisioned Ethereum as an improvement on Bitcoin, dealing with the latter's shortcomings and flaws. Like its older brother, Ethereum uses a payment network that is decentralized, has a proprietary cryptographic currency -- ether, and allows payments to be transferred through the Internet without requiring a bank or any other third party processor.

All transactions are recorded using a decentralized ledge, which is called the blockchain (more on this in another chapter). This ledger can be seen by anyone on the network.

What Is the Difference Between Ethereum and Bitcoin?

Being the next biggest cryptocurrency to Bitcoin, comparisons are inevitable between Ethereum and its big brother. There have also been rumors and claims of Ethereum being a 'bubble' because of its rapid rise in value and fame. But advocates of this fast-rising cryptocurrency state some advantages compared to Bitcoin that are worth taking a closer look at.

The first difference is that Ethereum allows cryptocurrency transaction records called 'blocks' to be created faster than it is currently done using Bitcoin. Although Bitcoin is now widely accepted by online merchants and is even being adopted by physical stores, the fans of Ethereum believe the efficiency of this cryptocurrency makes for better transactions.

The most important advantage of Ethereum over Bitcoin is that this cryptocurrency technology allows the computer applications to run on its network, not just ether by itself. The significant appeal of Bitcoin lies in the fact that it can't be controlled by any single party and it is not running via a central server. Ethereum improves on that by allowing not only the currency but other things as well to be run inside the network. Let's take Dropbox as an example. When you put files in this storage service in the cloud, you trust that Dropbox will take care of your files. But since Ethereum is in a decentralized storage, your faith lies on the people who are using the network and are interested in maintaining the network.

There are quite a number of applications currently being developed for Ethereum. The network is also used by startup companies in raising money through coin offerings. These offerings may exchange cryptocurrencies like ether for special access to a particular service.

One thing is for sure: Ethereum's popularity has skyrocketed and with it, the cryptocurrency's value. And it continues to grow just after two years of launch. All the ether currently in circulation now has a total value of 27.8 billion dollars, while it's 55.7 billion dollars for Bitcoin. Not bad for a two-year-old currency. And more investors are taking notice of this rising star.

What Affects Ethereum's Price?

Bitcoin is known for its steady and fast rise in value. Ethereum, however, took some time to pick up but when it did, the increase in value was significant. Climbing slower than Bitcoin did after launch, Ethereum held steady at approximately $10 during the first one and a half years. In March 2017, the cryptocurrency started climbing steadily then it hit $395 in June. It then plummeted back to $155 in July before starting to rise again.

The rise is greatly attributed to Bitcoin's popularity. When Bitcoin's value spiked, it also sparked a global interest in cryptocurrencies. Investors started looking for potential Bitcoin alternatives so Ethereum became a currency of interest to them -- hence, the sudden increase in its value. Ethereum also got some high profile support in the recent months coming from large enterprises which include JP Morgan, Microsoft, and Intel.

The significant drops in Ethereum's past are somewhat related to market wobbles, and general security concerns. These also include a deluge of fake news as well as sell-offs done by major investors. The 20 percent drop in Ethereum's value in July, for example, was fueled by a drop in investors' confidence due to a rumor that Buterin is dead. This was proven false later.

Who Created Ethereum?

This is also one of the notable differences between Ethereum and Bitcoin.

Who do you think created Bitcoin? Satoshi Nakamoto is a pseudonym that has always come up when journalists and users try to unmask the inventor of Bitcoin. It's true that Craig Wright publicly confirmed that he was the one who created Bitcoin, and the founding director of the Bitcoin foundation even backed his claim. But there is still some speculation regarding Nakamoto's real identity, with some people believing that there is not one person behind Bitcoin. Even advocates of the mega-cryptocurrency that is Bitcoin might not even know the mystery behind its creator or creators.

Information on Ethereum's inventor, however, has always been out there. Vitalik Buterin was born in Russia but he is of Canadian citizenship. Buterin has always been mathematically gifted even as a young child. He won the bronze medal at the 2012 International Olympiad in Informatics. At a tender age of 19, Buterin created the idea for Ethereum. It was 2013 and he spent the next couple of years building on that idea.

Buterin has been writing about Bitcoin in specialized websites since he was 17 years of age. He first got interested in cryptocurrencies when he heard about it from his father. In 2013, he came up with the idea of a software development platform and a decentralized mining network rolled into one. The Thiel Fellowship took notice of this proposal and granted Buterin a hundred thousand dollar grant in 2014. He quickly dropped out of college and with Joseph Lubin, a Canadian businessman, founded the Ethereum Switzerland GmbH in the same year.

Chapter 7: Smart Contracts From Ethereum

Ethereum is a practical example of smart contracts. In 2013, it was developed to create a nascent cryptocurrency technology. The reason behind this idea was to build upon the existing concepts, like Bitcoin, then improve the security and transactional speed.

Ethereum launched in June 2015 after an estimated $25 million was raised through crowdfunding in 2014. With its launch, it helped to create a new way in which cryptocurrency technology was being used, even though it was not a cryptocurrency like Bitcoin. Its intention was to create a program in which smart contracts can be built. Through blockchain technology, it is possible to have contracts that self-execute when particular terms or events are completed.

Ethereum is an example of a smart contract that is open source and that is also decentralized. The "ether" is used by

Ethereum to prompt peers who have the same network, to verify transactions, achieve consensus on what exists and what has happened, and also make the network safe – this will permit self-execution of the smart contract.

In Ethereum, smart contracts are run in the "Ethereum Virtual Machine" (EVM) which provides a more complete and meaningful coding language than what Bitcoin has for scripting. "Ether" is the native cryptocurrency that blockchain in Ethereum records the transfers.

A token, which can represent any asset, like a bond or house, can be traded using Ethereum, on a blockchain. Such a feat can help reduce the transaction times as well as the administrative costs, to disintermediate a couple of the already available service providers, and online markets that have developed recently.

Benefits of using Ethereum

• Tighter security with every participant, a server, and client at the same time.

The difference between Ethereum and other systems is that Ethereum has several server entities whereas, in other systems, they have a single server entity. This makes them vulnerable to exploitation by hackers and other potential attacks.

Ethereum is also resistant to hackers because it is decentralized, as it has zero downtime, even if there are sections of the system that are low. The integrity of data, when secured, verified and protected, makes the transaction log robust. Altering records is impossible as they can be accessed by anyone in the network, and they are also traceable. There are built-in balances, and checks to make sure that the accuracy of transactions is close to 100%.

• Ethereum is the best way to ensuring applications work properly. When Ethereum is used, with blockchain being the network behind the application, an order is executed, or a transaction is executed on its own, the output(s) are verified on their own, and the value between participants is distributed on its own. This makes it unnecessary to have different blockchains for each application, and also there is no need to have central administration processes used to monitor the processes.

DAO

In June 2016, there were some investors who had issues when they used Ethereum. The investors in DAO (a digital, decentralized autonomous organization), which is a kind of investor-directed venture capital fund, were exploited by hackers who discovered a weakness in the DAO code. One-third of DAO's funds were siphoned and sent to a subsidiary account, which was approximated to be worth $50 million. On 21 May 2016, the DAO's funds on Ether was approximately $50 million which was about 14% of the ether tokens that have been issued until June 2016.

$700 million was wiped off the book value by the hack of the Ethereum economy. The Ethereum Foundation, to instill confidence in the DAO investors and create an opportunity for their investment to be recovered, proposed a change in the underlying Ethereum code rules. The proposal was to have a constitutional amendment to freeze the account where the funds from DAO were being diverted. It was impossible to implement this solution, as it required those who were operating the computers using the distributed network system, to decide if it was possible to accept the changed code. It was determined if a majority of them agreed with the proposed solution, they would go forward with freezing the account.

There was a huge debate about whether the proposal should be adopted and how it would affect Ethereum's principle. In case the proposal was to be implemented, the bedrock Principle of smart contracts running as programmed, without the interference of a third-party, would have been downplayed. On the flip side, if the code was not implemented, the DAO would have collapsed, and the domino effect would have led to the disintegration of the confidence on Ethereum's platform.

The code was adopted in the long run and it made sure that the Ethereum blockchain restored the funds to the primary contract, making sure that the investors never lost their money. It was a controversial step indeed, but it led to a fork in the Ethereum, as the primary un-forked blockchain remained unchanged as Ethereum Classic; Ethereum broke into two different active cryptocurrencies.

This was a great way of testing investors and seeing if they wanted to be part of a decentralized economy, as there is no central authority to dictate sanctions and redress when there are problems that might occur.

Uses of Ethereum Smart contract in Banks.

Banks have had issues in the past with contracts that they form with their clients. Adopting Ethereum in their systems, through the use of smart contracts, can help them a lot.

According to an article in the Capgemini Consulting paper in October 2016, it states:

"Smart contracts, enabled by blockchain or distributed ledgers, have been held up as a cure for many of the problems associated with traditional financial contracts, which are simply not geared up for the digital age. Reliance on physical documents leads to delays, inefficiencies and increases exposures to errors and fraud. Financial intermediaries, while providing interoperability for the finance system and reducing risk, create overhead costs for and increase compliance requirements."

The statement above is a testimony of how banks can benefit when they integrate smart contracts into their operations. Some of the benefits include:

•	Administrative costs reduce considerably, and

- The burden of monitoring and verifying data is taken away from the banks' hands, and the smart contracts handle this.

The report by Capgemini does anticipate that banks who will be using smart contracts and distributed ledgers have the possibility of going mainstream "early in the 2020s."

Mortgages. Smart contracts could save a lot of money through the reduction of processing costs. The verification process involving all parties could be reduced to sharing access to electronic versions of the verified legal documents between those involved, and the information of the external source, such as title deeds and Land Registry records are accessible as well.

The costs saved are then passed to the client who could benefit immensely from the lending and interest rates, this, in turn, makes owning a home affordable. According to the Capgemini paper, it does estimate that, "consumers could potentially expect savings of $480 to $960 per loan," with banks being able to "cut costs in the range of $3 billion to $11 billion annually," when they lower processing costs in the primary process in the European and US markets.

Clearing settlement. The process of clearing and settlement with smart contracts is streamlined immensely. 40 global banks did try using smart contracts for this, and some are doing individual trial runs. The calculation of trade settlement amounts and managing approvals between parties and transferring funds automatically, when the transaction has been vetted and approved, is done by smart contracts.

An instance would be in 2015 when Depository Trust & Clearing Corporation was able to process more than $1.5 quadrillion worth of securities, which represented $345 million transactions.

Bonds. Complex computation can be done by blockchain, therefore smart contracts can be utilized in setting up and managing "smart bonds." Coding for creating such bonds will be in the permission area, which is, defining detailed rules on who is allowed and not allowed to hold the bond.

KYC. This is an expensive element where you onboard a new client, where each bank creates its own KYC. The result is a high-cost of client acquisition, and the customer will have to undergo a long process of opening a new account if it is at a new bank. One can incorporate this element in the blockchain. Information of the customers can be verified against the center records of blockchain network doing away with a third party.

When a client changes their address, smart contracts make it easy to alter the information, as compared to when administrative issues lead to delays. The coding in the smart contract will notify the clients immediately, and they will resubmit their proof, so that it is acceptable by the bank, without any manual intervention.

Problems with Smart contracts

1. Inflexibility – they cannot be easily modified as they are written as software programs. Once a problem arises, it is quite impossible to rectify it as the smart contract transaction needs to be completed before changes are made.

2. Contractual secrecy – traditional contractual documents have an NDA section where information shared needs to be kept a secret, but in smart contract, information is available to all parties involved, and an issue of confidentiality is likely to arise.

To resolve this, there are two ways it can be handled:

• Exploring the concept of "zero-knowledge proofs" to separate how verification of a transaction can be done without seeing its contents.

• Information shared through the use of advanced cryptographic structures that only the parties involved know how to access the information.

Legal jurisdiction – in case of an issue that requires the court or authority to intervene, it is impossible, as it is a decentralized distributed ledger network, meaning that there is no authoritative figure. With it being a new concept, very few courts are set up to acknowledge the legality of smart contracts.

For this to be a non-issue, the application of a smart contract transaction needs to be correct. Also, using simple terms in the contract would ensure transactions are carried out efficiently and correctly.

Chapter 8: What Is Ethereum?

So what exactly is Ethereum? After trudging through the sludge of the first chapter, I'm sure that you're keen to finally move on from the nonsense of arbitrary description and get to some practical knowledge. In this chapter, we're going to be doing what you came here to do: taking an in-depth look into Ethereum, what it is, what it can be used for, and what you have to gain from using it.

First thing's first: what is Ethereum? Ethereum is, at its core, just another kind of cryptocurrency. Any cryptocurrency aside from Bitcoin is often called an **altcoin**, short for **alternative coin**, which just means that the cryptocurrency is being used as an alternative to Bitcoin for one reason or another.

There are many reasons that one may choose to use an altcoin instead of Bitcoin. The first is that occasionally, Bitcoin's functionality can be a bit limited. The thing is that Bitcoin was the first of its kind. This meant that it developed as a cryptocurrency extremely quickly and picked up a lot of market share really fast. This means, also, that Bitcoin became the predominant cryptocurrency. This had some negative effects. When something is a legacy software - especially something in such dire need of stability as a currency - it becomes really difficult to make major overhauls and changes, especially when the currency is larger. This means that while Bitcoin does what it's supposed to perfectly well, it's unable to make larger changes. Perhaps **unable** is the incorrect word, but one way or another, the Bitcoin development team is highly hesitant to making large changes because so much is dependent upon the software at this point in time.

Another reason somewhat goes along with the first: at the end of the day, Bitcoin may be simply lacking in basic functionalities that other cryptocurrencies can accomplish perfectly well. For example, if one were looking to easily and efficiently make anonymous transactions without muddying the process with a bunch of additional applications or services like coin mixers, one could easily use Dash or Monero and have anonymous transactions built-in to their cryptocurrency. Likewise, Ethereum offers its own suite of unique services, which we'll be getting to in a momentarily.

Ethereum, ultimately, is just another cryptocurrency in much the same vein as Bitcoin. However, there are some key variances and technological innovations that make Ethereum a better choice than Bitcoin for many different instances.

Chapter 9: Why Ethereum Matters

Before getting into the basic concepts surrounding Ethereum and how to operate within this technology, it's important to understand why Ethereum is important and what it means for society as a whole. As with most inventions and innovations, Ethereum was born out of a societal need that was not being met. This chapter will briefly discuss that need and will also touch on blockchain technology as a whole, it being the technology enabling platforms like Ethereum to function. We will end this chapter discussing how Ethereum differs from other types of popular cryptocurrencies, as this will pave the way for conceptualizing the following concepts presented in this book.

Eroded Trust in Banking Institutions

Prior to the 2008 economic collapse, people generally had trust in their banks. The privatized structure of a bank largely allowed bank consumers to feel as if their money was being handled with both respect and transparency. When the economy fell and the banks were ousted as being filled with employees who were being enticed to hide information from their customers, it suddenly became clear that the banks did not entirely deserve the trust the public had given them. While poor and deceptive investment decisions (especially in the housing market) largely led to the economic collapse of 2008, the takeaway feeling that rippled through the public after thousands of homeowners went into foreclosure was that their banks had lied to them. This prompted the public to take back the trust it had given these banks. It's at this point when cryptocurrency made its debut.

The Emergence of Ethereum

While bitcoin was the first official cryptocurrency to make its way to the digital marketplace, Ethereum was actually created by a man who was working for bitcoin when it was first developed. His name was Vitalik Buterin. Bitcoin was developed in 2009, and Ethereum was not officially launched until early 2014. As you can see, Buterin had quite a bit of time to fully understand the blockchain technology behind Bitcoin, prior to launching Ethereum. As of June 2017, Ethereum has netted over 400 million dollars in profit, and it's also notable that multiple foreign entities have endorsed the platform. Both Japan and Russia have publicly endorsed Ethereum's capabilities. These international entities are a primary reason why the popularity of the Ethereum has increased so rapidly.

Decentralization and Blockchain

Without going into too many specifics about how blockchain technology works, it's important to understand that Ethereum and other cryptocurrencies like it are all supported by blockchain technology. At its core, blockchain technology enables untraditional banking systems to exist through the notion of decentralization. In other words, when it comes to traditional monetary transactions, a bank is considered to be a third party who is involved with making sure that two people exchange their money safely. For example, if you are looking to give your friend Joe money to cover the cost of a camping trip you planned, you may decide to give him this money you owe him in the form of a check. Without the bank's help, this money that this check represents would not be able to be transferred into money in Joe's account. The bank processes the check for you, and makes this transaction with your friend possible.

With cryptocurrencies, you don't need to use a bank. Instead of trusting one bank or single institution with your money, you instead trust a network of multiple authoritative and mathematically reinforced administrators. All of these administrators must agree that your transaction is correct prior to it being approved. This is what lies at the heart of the blockchain technology. By dispersing the trust that's typically involved when you transfer money through a bank, more transparency is possible. Not only that, but a typical blockchain application will also provide you with a ledger. This ledger is able to keep track of every single transaction that occurs within the blockchain system. This way, everything that occurs within the blockchain is documented and can be easily found if and when needed.

Beyond Counting Coins

One of the biggest reasons why Ethereum is becoming arguably one of the most reputable forms of cryptocurrency on the market right now is because its platform goes beyond currency. Trust in a marketplace can extend from currency to mortgage ownership to marital status. In other words, non-monetary transactions are the primary transactions that Ethereum targets. While Ethereum does have its own type of cryptocurrency known as Ether, it uses what are known as Smart Contracts to complete and solidify contractual agreements between parties without any physical human intervention. By setting up the contract parameters within the Smart Contract, the person signing the legal document is then bound to the terms within it. It's important to understand that even though Ethereum differs from other types of cryptocurrency platforms that currently exist, all of these cryptocurrencies are operated through blockchain technology.

This expansion of blockchain to include contractual agreements between parties is a major reason why Ethereum is currently receiving so much attention. From a corporate perspective, being able to send someone a document to fill out rather than formally meet with someone will save money because these companies will likely have less employees to hire. With less people to hire, there is much less room for human error. Additionally, with less room for error, greater efficiency and accuracy is possible. Lastly, unlike with bitcoin, Ethereum allows you to actually create your own blockchain networks. This means that you are able to develop your own currency, your own terms of ownership, and your own contractual obligation requirements. Ethereum is thus much more adaptable to an individual's ownership preferences and requirements when it comes to running a business or operating a network for a small niche of people. Bitcoin, contrastingly, was the first blockchain network to successfully function, but is only capable of allowing people to trade funds.

At this point, you should have a working understanding of how Ethereum came to emerge within a society where trust in the marketplace was at an all-time low. It can also be argued that the emergence of

Bitcoin ultimately led to the introduction of Ethereum as a concept and digital tool. This book will get into more detail regarding how Bitcoin and Ethereum differ, but it's important to understand that the question of whether Bitcoin or Ethereum will outperform the other is still largely unanswered. A primary reason why cryptocurrencies are so exciting is because the future of this technology is still largely unknown. Questions regarding security, duplicity, and accessibility still need to be figured out. It seems that the only way to truly answer these questions is to wait until the boundaries that surround cryptocurrency in general are tested.

Chapter 10: Understanding Cryptocurrencies

What is a cryptocurrency?

To understand cryptocurrency, it is important to appreciate that it is made of 2 words i.e. crypto and currency. I know that you understand what a currency is i.e. a medium of exchange i.e. money. What of crypto? Well, crypto comes from the fact that the currency works on the principles of cryptography to store and transmit data securely.

This means it is safe to define cryptocurrency as a digital currency whose purpose of creation is to exchange digital information through a process that is enabled by certain principles of cryptography. Cryptography refers to a technique that is used to anonymously store and transmit data in specific forms so that the person it's intended for is the only one who can read and process it—to verify and secure transactions, and regulate the creation of new units of a specific cryptocurrency.

You can also define cryptocurrency as an encrypted digital currency that is decentralized and that can be transferred between peers and is often confirmed in a public ledger through a process referred to as mining.

As a complete beginner to cryptocurrency, what you need to know is that cryptocurrency is just like a debit card or PayPal with the only difference being that the numbers on the screen actually represent the cryptocurrency as opposed to a fiat currency like the dollar. In fact, to use cryptocurrency, you really don't have to understand every bit of it (the same way you really don't have to understand the intricacies behind the functioning PayPal and credit cards to use them.

Nonetheless, this doesn't mean you shouldn't seek to understand the workings of cryptocurrency even if a little. What you should seek to understand should be the concept of digital currency, cryptography and blockchain (both the technology and the ledger of transactions).

Why is that?

This is because cryptocurrency is essentially type of digital currency in which transactions are usually recorded on a type of digital ledger referred to as blockchain with the whole process being secured by cryptography!

Cryptocurrency works very much like bank credit on any debit card given that in each of these two cases, there is a complicated system responsible for issuing currency, which is also responsible for recording balances and transactions behind the scenes to enable different people to receive and send money electronically. However, there is a slight difference; with cryptocurrency, instead of the bank or government issuing the currency as fiat currency and maintaining the ledgers, with cryptocurrency, this process is managed by an algorism.

This kind of currency doesn't exist in physical form like fiat currency does; it exists in computers, which essentially means that transfers can only be done from computer to another between peers but the interesting bit about it is that there are no middlemen like the bank. Instead, the transactions are usually recorded in what's referred to a blockchain, a digital public ledger. This ledger along with the transactions are then encrypted with the use of cryptography. Cryptocurrencies are decentralized i.e. they are controlled by the users as well as a computer algorithm- as opposed to a central issuer like the government.

So how do cryptocurrencies work? Let's discuss that next.

Chapter 11: Ethereum Use Cases

The Ethereum technology is still in an infancy stage. Therefore, any Company and technology enthusiasts that are interested in

staying current on the implications of the technology must always track both technology and cases of its use in the industries.

Below are some of Ethereum's use instances:

· Smart Contracts

· DAOs

· Dapps

Let's jump in and explore those use cases.

Bright ContractsBright contracts are self-executing algorithmic Codes that are stored and replicated on a dispersed ledger

system--the Blockchain. The algorithm is executed by means of a network of nodes that run the Blockchain and can lead to routine

updates of the ledger. To put it differently, a smart contract is a software that executes on the Blockchain if an activity is

triggered.

For the application to operate, it has to be confirmed By many nodes in the distributed community to ensure that it's trustworthy.

In case you were considering the Blockchain technology and its own power of spread reliable storage, then intelligent contracts

have the capability to provide trusted computations on the dispersed storage.

With intelligent contracts, there's no requirement for a Single source of management. Bright contracts use the Blockchain

technology where several parties-- autonomous computers--utilize consensus mechanism to constantly check and re-verify any updates

to the ledger. This promotes transparency.

Since all of the nodes at the Blockchain network Are running the exact same code, with every verifying the other, smart contracts

will be visible to all. Any node can start looking into a wise contract, and if it is pleased with the logic, it can use it. On

the flip side, if the node doesn't agree with all the code, then it doesn't run it. That's how transparency is promoted in

contracts that are smart.

Smart contracts can provide benefits for a Broad range of businesses such as banks, healthcare providers, and insurance

businesses. When employed correctly, these associations may benefit from reduced risks, real-time processing, precise and verified transactions, fewer third-parties and reduced costs.

Smart contracts can be deployed using the Solidity language or Pyethereum scripting languages. Smart contracts programmatically

execute only when they get instructions, which can be in the form of a transaction in the EOA. The contracts may either push or

pull funds and request these activities from different contracts while calling on the contract code to perform activities that are

dynamic. Here's a sample of a smart contract code:

pragma solidity ^0.4.0;

Contract MySimpleStorage

uint storedData;

function Place (uint mydata)

StoredData = mydata;

function Get () continuous returns (uint)

return storedData;

DAOsWhile smart contracts on their own are Fascinating, it is the notion of a huge number of unified contracts operating

collectively that showcases the extensiveness and potential worth of Ethereum's technology. Combined with DAOs (Distributed

Autonomous organization) or DACs (Distributed Autonomous Corporation), smart contracts can promote enforcement of principles in

the ecosystem.

Here is the way the DAOs work:

· A group of nodes writes the smart contracts (codes) that will run the company or organization.

· There's the first funding period, where nodes add funds to the DAO by buying the components that represent ownership (this is called a audience sale or Initial Coin Offering (ICO)) to supply it with the funds it requires.

· After the funding period ends, the DAO starts to function.

· Nodes can then make the suggestions to the DAO on how to spend the money, and the nodes which have been purchased in will vote

to approve the proposals.

At the moment, all the Intelligent contracts which we Have discussed are owned and implemented by other reports which we presume

were humans. But there is not any bias against the robots or other folks in the Ethereum ecosystem. In particular, the wise

contracts may create random actions just as with any other account could.

They can possess tokens, participate in the audience Earnings, and also act as voting members of different contracts. The DAOs can facilitate such arrangements in the Ethereum ecosystem. The way this particular democracy function is that any wise contract code must have an owner that functions as the administrator.

The Owner can add (or remove) voting Members in the organization. Any member may create a proposal to the ecosystem, which is in the form of a contract trade to either ship the Ether or perform a few smart contract. The members can then vote to support or reject the proposal. After a predetermined period is selected and adequate members have voted, the contract counts the amount of votes and if there are enough, it is going to execute the given transaction.

DappsThe Principal objective of the Ethereum system Is to function as a platform for the development of Distributed applications

(Dapps). Dapps can be developed from one DAO or a succession of DAOs that work in unison to create an application. This can lead

to something very similar to apps such as Google Chrome or Microsoft Outlook.

Such apps can be made to reach a certain functionality. However, for an application to be considered a Dapp, it must meet the

following criteria:

· It must be wholly open-source and operate autonomously without a entity managing the majority of its tokens. The app may adapt

its protocol in response to the planned improvements and promote the feedback but all changes have to be determined by consensus

of its users.

· Its records and data of performance must be cryptographically saved in people. Blockchain should also be distributed to prevent

any fundamental points of failure.

· It has to utilize cryptographic tokens to get the program, and any contribution of value from the miners ought to be rewarded in

app tokens.

· It has to generate tokens according to some standard cryptographic algorithm that acts as evidence of their value.

Dapps are grouped into four categories:

· Smart contract services, utilities, and analytics

· Information validation and Oracle solutions

· Gambling and games

· Registry and corporate governance.

Chapter 12: Who Should Invest In Ethereum And Why?

It is easy to see how Ethereum's prices have consistently increased since it was first introduced in 2016. Some reports are even showing returns as high as 3000%. Those kinds of reports are more than enough to stir up excitement about investment opportunities. However, like anything else, investing in Ethereum is not the best choice for everyone. There are several questions you must answer about Ethereum and about your own personal investment style before you can determine whether or not it is the right investment choice for you.

Let's first talk about Ethereum's potential in the cryptocurrency marketplace. One of the first things you want to determine is its path for the future. What things indicate that Ethereum will be around in the next year, five years, or even twenty years?

If you have been following the reports you probably already know about the Enterprise Ethereum Alliance (EEA), an organization set up specifically to connect Ethereum with many major industry leaders. Already large corporations like Microsoft, Santander, JP Morgan, Credit Suisse, and Brainbot Technologies are on board. They've partnered with the currency primarily because of its multi-purpose design that can be adapted to just about any industry.

These partnerships have also been an impetus for the rising value of the coin and are a major cause of the bullish trend we have been witnessing in recent months. According to many cryptocurrency experts, it is very reasonable to see the value of Ether continue to rise, especially as more and more larger corporations begin to utilize its capabilities.

Prediction for Ether's price predictions has also been very promising. According to many experts, if history is to repeat itself, the price of Ethereum is expected to increase to more than $20,000 per coin within the next year and even more beyond that. If you're looking for a long-term investment option then now might be the best time to get in; the prices are still low enough so you won't have to put up a significant amount of cash. You'd actually be investing at a low price before the bullish trends begin to rise. At the time of this writing, Ethereum's investment stats look extremely promising.

Average Daily ROI 1.69%

Average Monthly ROI 66.11%

Total ROI 34,515.80%

Volatility trailing 30 days 43.44%

Volatility trailing 7 days 20.98%

If its track record continues like this, it is a strong possibility that Ethereum will be around for quite some time. For the moment, it appears to be an ideal investment opportunity for someone looking for a long-term situation. But the next question to answer is if it is the right opportunity for you.

Is Ethereum the Right Opportunity for You?

While there are always a lot of statistics and logic involved in making an investment decision, there is also the need to personalize the data and tailor it to your personal needs. Just as there are numerous investment opportunities, there are also more than a few investment styles and needs. The decision to invest in anything will also depend on the individual needs of the investor. To determine whether or not Ethereum or any other type of cryptocurrency is right for you will depend on where you fit in the whole scheme of things. To determine this, you need to understand your personal level of risk tolerance, your investment style, and create your own investment plan.

How to Analyze Your Risk tolerance

Risk tolerance is a term often heard but never fully understood, especially for a new investor. We often hear references to investment opportunities that vary based on a person's risk tolerance, but we first need to fully comprehend what it is before we can determine where we lie on the spectrum.

Several factors must be considered when determining your risk tolerance. One has to do with how much time you have to invest. It is a general course of wisdom that says that the younger you are, the more time you have to wait for your returns to come in. While many people think the returns they want to achieve when they invest they also have to consider the amount of time they have. If you are a young person, perhaps in your twenties then you would have a long-term investment horizon, but as you age, your investment horizon begins to shorten.

This knowledge will determine if you would want to invest in Ethereum or not. If you have a short investment horizon, you're probably going to be looking at opportunities that will require you to see results very quickly. Ask yourself...when will the returns be needed? If you need them quickly, you would want to take a more conservative approach to the investment. If your investment horizon is longer, you could probably afford to take a riskier stance with Ethereum, perhaps even trying some day trading or margin trading.

You also want to think about your risk capital. What is your net worth? Your net worth is simply how much money you have to spare. Simply list all of your assets and then subtract the total amount of your liabilities. This will tell you how much capital you are in a position to risk. Those with a high net worth will be able to take on more risk and can afford to take a more aggressive stance when it comes to investment choices, but those with a low-risk capital will need to be able to play it safe, so they don't lose too much and end up causing financial hardship.

Be careful not to fall into the trap that many investment opportunities present. It is easy to be drawn into riskier investment measures when you have a small amount of risk capital to work with. The idea of fast money or the opportunity to win big can be a powerful lure, but if you have only a little to invest, it is better to play it safe and build up your portfolio slowly. Once you've increased your risk capital, then you can venture into more dangerous waters for those bigger returns.

Next, you want to think about your ultimate goals. What do you want to get out of this decision? Are you setting up a plan for your retirement? Are you planning an around the world cruise? Or do you just want to build a nest egg that you can tap into later? Your goals will also factor in helping you determine how much of your money you are willing to risk. If you're saving for your child's college education, you'll probably be more protective of your money than someone who is planning an around the world cruise.

Finally, you want to think about your experience in investing, especially when it concerns cryptocurrencies. If you are new to this type of investment, you should proceed with caution. As you learn more about the market and how it moves you could take more chances with your money.

Investing in Ethereum is very much like investing in anything else. There are levels of investment risks that everyone should take. But once you know your risk tolerance the easier it will be to decide if investing in Ethereum is right for you and if it is, then how much of your money you are willing to risk to get those returns.

Knowing your risk tolerance is not just about the anxiety that naturally comes with a volatile market like cryptocurrency. It is about being able to balance your financial position and your goals and finding the right place for you to enter the market and with how much.

Determining Your Investment Style

Now that you know your level of risk tolerance, you need to determine what type of investor you want to be. You have probably already realized that understanding your risk tolerance has already allowed you to remove certain investment options you were already thinking about. Now, it's time to narrow that field down just a little bit more.

People tend to get very excited about investment tools like Ethereum. They see the meteoric rise in prices and they are drawn to them like a string pulling at their hearts. But when it comes to serious investing, that is exactly what shouldn't happen. To be successful in this type of move, it is important to approach it logically rather than with the fever pitch of adrenaline pumping through your veins.

Risk tolerance determines how much of your assets you are willing to risk to get a return on your investment, but your investment style reveals to what degree you want to delve into the market. There are at least three different major investment styles where most people fit. Of course, these can be broken down into smaller degrees so you might find yourself a blend of two but in determining where you fit it is important to understand these basic investment styles.

Passive: As an investor, you should first think about to what extent you trust the advice of the experts in telling you how to get the largest returns.

Those who want to lean entirely upon the counsel of professional advisors, and only follow the direction on when to get in or out of the market, or whether or not they should have a long or short-term investment would be considered to have a passive investment style.

These people have full confidence in the words of the experts believing that they have their finger on the pulse of the market and are in a better position to predict future movements.

Active: There are also those investors who want to do all the legwork themselves. They don't really trust the advice of others who may not know their personal situation. They make their decisions based on their own research and calculations.

Growth: Are you looking for a coin that is on the fast track and will grow exponentially over the coming months or are you willing to sit back and wait for a slower but more stable return on your investment? Those who are looking for a faster rate of growth will be interested in coins that are already showing faster growth rates and higher profit margins.

Value: The value investment style focuses on the price. They want to get in at a low price and sell at a higher price. The rate of growth is not necessarily that important. These people more than likely lean towards a long-term investment strategy because they are not focused on how fast the market moves; their only concern is that the market is moving in the right direction.

Capitalization: When it comes to stock investing, those companies with a small cap usually bring a better return on their investment because there is more room for growth. However, when it comes to cryptocurrencies, that is not the case. Small-cap coins are often more vulnerable to attacks because buyers can more easily manipulate the market. When it comes to investment style, any coin that is under one million in market cap is already in the danger zone, making them a much riskier investment.

The good news is that Ethereum is not a small cap coin, so while there are risks involved, they are in a much better position than many other coins.

It is very important that you determine your personal investment style before you decide to enter the market. It will help you to determine not just if you're going to invest in Ethereum but how much you plan to invest, and it may even help you to pinpoint a comfortable time to take the plunge.

Creating an Investment Plan

Now that you already know your risk tolerance and your investment style, it's time to create your plan. Whether you're a first-time investor or you've already dabbled with other coins, it is important to lay out on paper your expectations and milestones you expect to achieve.

You've probably already heard about the many rags to riches stories that are floating around the Internet, and many of them are true. However, those stories are not the norm for most people. While it is possible to achieve this type of success with Ethereum, it should not be without some sort of game plan.

By the very nature of the cryptocurrency market, it is never wise to invest all of your assets into one coin, even one as good as Ethereum. For that reason, you'll need to determine what percentage of your assets you want to put into Ethereum and what percentage you want to put in other coins.

Again, go back to your risk tolerance and try to determine what other coins will be able to support your acceptable level of risk and matches your investment style to decide which ones will be best designed to work in tandem with Ethereum. Many experts recommend that newcomers take a more conservative approach while experienced ones who know the market can afford to take a higher risk and dabble in something a little more dangerous.

Those with a low horizon will likely want to lean towards balancing their portfolio with more traditional and proven coins and those with a long horizon will more than likely lean towards some of the newer coins or ICOs.

Getting Ready for the Big Day

Before you can buy your first Ether, it is necessary for you to get prepared. Buying Ethereum is not the same as buying anything else. Certain things must be in place before you begin.

The Wallet

Because Ethereum is a digital currency, you will need to have a digital wallet to store it. Like any other wallet you might have, a digital wallet is primarily used as a depository for your cryptocurrency. However, it has other functions too that you will have to rely on from time to time.

Before you can begin to understand the role of the wallet, you have to grasp the concept of digital currency. Since cryptocurrency does not have any physical characteristics, your currency will exist in the wallet in the form of codes and numbers. What your wallet will actually hold are the private keys that give you access to those coins.

Every wallet comes with at least two keys (some have more) a public and a private key. The public keys allow the holder to send a portion of their currency to another user while the private key is to make sure that the transaction remains secure. Anyone who holds the private key has access to your coins.

This means that just as you wouldn't leave your physical wallet laying around for anyone to pick up, you must use the same diligence in protecting the keys to your digital wallet.

There is a wide variety of wallet options to choose from. The one you choose will depend on how you plan to use your Ether.

Software wallet: A software wallet is a program installed on your computer. You have total control of the wallet, and as long as you protect your private key, no one else can have access to your currency. Simply choose the one with the features you like and download it into your computer and you are ready to begin using it.

This type of wallet is about as secure as you can get but there are some vulnerabilities. They are only as safe as your computer. Because they are connected to the Internet they may be affected by malware and other cyber issues, so they are not always the safest in terms of protecting your currency.

Paper Wallet: Paper wallets are a means of protecting your private and public keys in a physical form. Once you have received your currency, simply print out the keys and store them someplace safe that is offline. As long as the data stored on the computer is deleted after you print out the details your money will remain safe.

The advantage of paper wallets is that all access to your currency is stored completely offline making it inaccessible to cybercriminals who may be looking for ways to access it. The drawback is that they are more exposed to environmental issues that could destroy the coin. By keeping the paper in a secure location where it can't be damaged by water, fire, or other environmental hazards you can be sure that your currency remains safe.

Web Wallets: Many exchanges offer web wallets to make it convenient to access your currency when you need it. Setting up a web wallet with the exchange you plan to trade with can be very convenient. Any earnings you make from the sale of coins are automatically deposited into your web wallet, and anytime you want to make a purchase, the money is drawn from your wallet balance just like when you use an ATM or debit card.

Obviously, convenience is the key advantage of a web wallet. With just a few keystrokes you can buy, sell, trade without any hassle. Using software or paper wallets means transferring the funds from your offline wallet to the exchange, which could take time. Considering the volatility of the cryptocurrency market, there are times when you have a very short window of opportunity to make a good deal.

Hardware Wallet: These wallets work on the same level as a USB port. They are small portable devices that can be inserted into your computer when you are ready to make a trade and removed when your transaction is completed.

Advantages of the hardware wallet are the convenience. Once your currency is stored, all you need to do is insert the wallet into a computer device hooked up to the Internet and perform whatever transaction you want to do. When completed, disengage your wallet and store it for the next time.

Exchange Wallets: Many new investors often wonder why they can't leave their coins on the exchange where they are trading. You automatically receive an exchange wallet when you open an account. However, there are some obvious risks associated with these wallets.

Not only do you not have control of your keys when you hold currency in an exchange account wallet, but they offer a third party service, and you are entrusting them with the responsibility of keeping your money safe. In addition, they are usually the primary targets for hackers and other types of cybercriminals looking to steal away with your coins. Without the structure of regulation behind them, they are extremely vulnerable on a number of levels. Ideally, you want to keep your coins on a platform that you have complete control over.

Generally, it is recommended to have at least two wallets, one hot and one cold. Hot wallets are those that you will use to buy, sell or trade with while cold wallets are those where you will store the bulk of your coins. These are kept offline where they are the safest, so there is a much lower risk of tampering or loss.

Once you've made the decision to invest in Ethereum, you should find the wallet you want to use. This needs to be done so that you have some place to deposit your funds when you make your first transaction.

For Ethereum, many people recommend the Mist Ethereum wallet. It is considered the official wallet for Ethereum. It is not the only option you have, but it has many characteristics that are ideal for the Ethereum investor.

Access to your own private keys

Easy to use

A strong community backing it (if you have problems)

It can be backed up and restored if your passwords or keys are lost.

It is compatible with a number of different systems.

No matter which wallet you choose, you need to make sure that first, it is compatible with the computer you'll be using and 2) you have enough storage space to store it. The Mist wallet, Ethereum's official wallet, for example, takes up almost 10 megabytes of space. It will come in a zip file, and you'll need to unzip it and install it on your system.

Once the wallet is installed, your next step will be to launch it on your computer and sync it with the Ethereum network. It would be a good idea to find some detailed information about the wallet before you begin, as this will definitely simplify the process of getting set up. As a matter of fact, there are a number of

YouTube videos that will walk you through the process painlessly.

Choose an Exchange

Now that your wallet is set up, you can start buying and selling Ether. However, you will need to know how to generate your wallet address so that the exchange knows where to deposit your Ether.

The good news is that because Ethereum is the number two coin on the cryptocurrency market, it will be easy to find an exchange that will allow you to trade them. Several factors must be considered when choosing the exchange. Just because most exchanges may allow you to trade Ether, doesn't mean they are all the same. Do careful research to find one that best suits your needs. Here are a few things you need to look for.

Fees: one of the first things you need to find out are the fees the exchange will charge. It is nice to think that they'll just let you buy and sell for free but all exchanges work on a commission. Some will charge a fee for every possible use you have while others will not. You don't want to end up paying all of your profits to the exchange in fees.

Location: Depending on the location of the exchange you may find that the way they do business will vary. Exchanges found in the western countries like the United States and Canada are often subjected to governmental regulations to protect your investment, but those found in eastern countries like China may have no regulations they are compelled to comply with. If anything were to happen, your money might be easily lost.

Customer service: Some exchanges only have customer support via email while others may offer direct human-to-human chats. It is also good to find out if their customer support is in a number of languages. It could be difficult to work out a problem when no one can speak in a language you understand.

Security: Find out exactly what security measures they have in place. The risk of losing your currency through cybercrime is high. Since most exchanges store their data in the cloud, if you choose to leave your money in your exchange wallet if something were to happen it could be impossible for you to recover if they don't have some sort of security measure in place. Exchanges are not insured with FDIC like banks and other types of financial institutions so once your currency is lost, or if the exchange folds, you may lose everything.

These are just some of the things you might want to investigate before you choose an exchange to work with. Chances are you'll think of more once you start interacting with them.

After you've decided on your exchange, you need to setup your account. Hopefully, the one you choose will be very user-friendly so you can navigate their system easily.

For newbies, an online wallet is the easiest way to get started. It provides you with an address for the exchange to send your currency to, but many are able to handle more than one currency. With online wallets, you can actually pay for other coins with crypto rather than exchanging your crypto for dollars and then use that money to buy a new coin. It not only will save additional steps but it can also save in fees too.

To open your exchange account, you will need to go through the signup process, which should be pretty straightforward. Depending on how much you want to invest, to begin with, they may ask for identification or other personal information. Once you've input all of the requested information, they will then ask you to go through a verification process; usually by sending a photograph of your identification, or uploading a picture.

Some exchanges will do this automatically, while others may have you wait for a day or two until they validate the information you've given them. So, if you're anxious about getting started with Ethereum, it is a good idea to set up your wallet and exchange well before you are ready to enter the market. There is nothing more frustrating than seeing the price right where you want it to be, but you can start trading because you're not set up yet.

It is important to remember that while the setup of a new account is pretty basic stuff, you should never forget your personal security measures when you're exchanging personal information online. It is true that exchanges get hacked from time to time, but the majority of cryptocurrency losses are usually the result of lack of care by the user. Protect your email account, your passwords, and any other security measures that are already in place. When you do this, you can greatly minimize the risk of your account being exposed or vulnerable.

Hopefully, you've picked an exchange that uses two-factor authentication, so it gives you an additional layer of security. Some points to remember as you go through the process.

Use your real name and address

Use your real phone number

Choose a password that is unique. Don't settle for a password with five or six characters. Some exchanges require passwords that are at least 32 characters long that are made up of a combination of upper and lower case letters, along with numbers and symbols. The more complex the password, the safer your investment will be.

While every exchange is different, you can be expected to provide the same information, no matter which one you choose. The good news is that once all your information is verified and you've gone through the process, you are ready to start investing in your Ethereum.

Even though you plan to start with one wallet, in time you will likely find that you will need more than one (hot and cold) and you may find that you need to deal with more than one exchange. As long as you remember to get setup long before you need them, you should be fine.

Chapter 13: The Blockchain

Cryptocurrencies require systems and structures that enable them operate effectively and securely. At the core of any successful cryptocurrency is the blockchain. It is an important platform that ensures all transactions are decentralized and secure. But what is the blockchain?

The blockchain can be defined as a decentralized, digital public ledger where all cryptocurrency transactions are recorded. Whenever transactions occur, they have to be registered, confirmed and authenticated. Each transaction is then recorded onto an individual block. This block, once confirmed, is then added to the blockchain.

Therefore the blockchain is a dynamic public ledger that keeps growing all the time. Users are able to access the blockchain at anytime and confirm transactions. On the blockchain, blocks are secured and linked using the latest cryptograph technology.

They are added chronological order which means as soon as a block is confirmed it is added to the system. The block contains what is known as a hash pointer. This hash points to the previous block. The block contains transaction information and a timestamp.

Blockchain are designed to resist any modification of data. This is important because miners confirm cryptocurrency transactions before entering them into the system. To maintain integrity of the system, the information should remain unchanged.

In its simplest form, the blockchain is simply an ever-growing digital ledger with blocks of information added periodically. However, the blockchain can be adopted for use as a distributed ledger. This is where it can be adopted for use by cryptocurrencies. In such instances, the blockchain operates as a peer-to-peer network. Once blocks are added to the using the protocol prescribed, data they contain cannot be altered. If alteration is to occur then it will affect all subsequent blocks and this is a decision that will require the support of a majority of users within the network.

The initial blockchain was designed and developed to support the first cryptocurrency, Bitcoin. The technology adopted for use is referred to as DLT or distributed ledger technology. This technology on the blockchain is finding application in many different sectors. Primarily, though, it is used for processing, entering and confirming transactions within cryptocurrencies systems.

Apart from confirming transactions, the blockchain can be programmed such that it can accept any document and any data entered becomes an indelible record that can be authenticated and verified by users but cannot be altered. In any case, should alterations become necessary, then the entire community will have to give consent and not just a single, centralized entity.

Blockchains serve as a great example of distributed computing systems that are extremely secure by design. All these unique features make the blockchain an ideal platform for keeping accurate and incorruptible data such as medical and dental records, motor vehicle and firearms registration details, identity management, land titles and records management and so much more.

The technology behind the blockchain

The blockchain is a peer-to-peer network that is one above the Internet. It is defined as a digital and distributed ledger with a distributed database. Decentralization is a key feature of this technology. Another outstanding feature of the blockchain is that the database containing all information, blocks and data exists across different computers within the network.

The blockchain network consists of computers which are referred to as nodes.

The nodes are interconnected in what is referred to as a peer-to-peer network. This kind of network eliminates the need of a centralized computer server.

Most organizations across the globe still use centralized systems with a database and server computer. The server then becomes a lucrative target for hackers and malicious attackers. On the blockchain, it is very difficult for hackers to penetrate because all the data is decentralized, encrypted and made available to users.

Every time there is a transaction on the network such as a purchase, a payment and so on, the transaction details have to be recorded on all nodes, or computers, within the entire network. Therefore, all members of the network, commonly referred to as users, get access to updated information. The information cannot be deleted or altered by anyone. In the end, there is a single, verifiable ledger with accurate information rather than a couple of ledgers with conflicting data.

A brief history of the blockchain

The blockchain first hit the headlines in 2009 when Bitcoin came to life. It can be found in the initial bitcoin code. When Satoshi Nakamoto released a white paper detailing the world's first cryptocurrency, the world got to see first hand the blockchain in application.

Satoshi Nakamoto, a pseudonym for the creators of bitcoin, released the code to the public via the Internet. Satoshi also mined the initial cryptocurrencies or bitcoins for the platform. The initial block that was mined is referred to as the genesis block. By mining the first bitcoins, they introduced the world to its first cryptocurrency.

The world's attention was attracted to bitcoin, especially after 2013. Everyone wanted to buy some bitcoins and so interest grew. All these people who purchased or sold bitcoins transacted on the blockchain. It is an integral component of bitcoin.

Since then, there have been more and more cryptocurrencies and all are based on the blockchain. Many other institutions and organizations have expressed an interest in the blockchain. In summary, the blockchain is a web-based protocol that finds application with DLT or distributed ledger technologies. Experts in the tech industry claim it is the biggest invention since the creation of the internet.

Digital signatures

Each transaction on the blockchain contains a digital signature. This signature is entered using public key cryptography. This technology uses two keys, one which is public and the other is private to endorse transactions. All members of the network or all the nodes within the blockchain have access to this information and can thereby check, confirm or verify any transaction.

However, only information based on the public key is available to everyone on the network. The information on the private key is only available to the receiver so they can decipher the message and check the details. The keys are also used to confirm the identity of a user.

Blocks of information

The blockchain derives its name from the many blocks it consists of. A single block contains information pertaining to a number of transactions. There is additional information on the block as well. This information includes a time-stamp, a reference to a previous or parent block, proof of work and a header. The header has 3 pieces of information including the time-stamp, proof-of-work and transaction details.

Mining of cryptocurrencies

All cryptocurrencies have to be mined into existence. Mining can be defined as the process of creating new cryptocurrencies or the process of adding new blocks to the blockchain.

Different cryptocurrencies have different mining speeds. On Bitcoin platform, it takes 10 minutes to mine one block. On Ethereum, it takes much less time. Therefore, it takes 10 minutes or less to complete a Bitcoin transaction.

Every new transaction that is added to the blockchain has to be validated. Mining is the process through which validation is accomplished. A miner, who is more often a computer programmer or other expert, is required to solve a complex mathematical problem. This process requires enormous computing power and consumes both time and electricity.

The complexity of the mathematical problems keeps increasing and this demands increased computer power. Miners are rewarded with a share of the cryptocurrencies that they mine and also get to share in some of the transaction fees that are charged on the platform.

Uses of the blockchain

The potential to transform different sectors of the economy through blockchain technology is immense. This is because the blockchain functions more as a foundation technology that can be molded, adopted, and programmed into a more specific application. It has the potential to design a foundation for social and economic systems that can be adopted across the globe.

This approach of the blockchain to provide solutions for systems is different from the current traditional approach where business models are programmed largely to provide lower cost solutions that eventually overtake firms traditional models. However, even with such revolutionary technology and proof of concept, there are few systems developed out of blockchain. This is due to the fact that blockchain is still new technology and there are plenty of products still being developed.

The blockchain applications may be disruptive in nature and may affect the way current models work. They can be adopted into multiple areas so as to enable business to adopt modern methods of processing payments and securing transactions. Blockchains eliminate the need for intermediaries and cut out middlemen. Blockchains also get rid of service providers. They instead provide protocols for processes such as digital transactions and use of virtual currencies. All these enable businesses to cut costs and save on otherwise expensive processes.

The digital ledger property of blockchain helps eliminate risks such as fraud, theft and so on. Using applications with this property will remove the need for a trust service provider. This means fewer resources will be tied up in solving disputes. Also risks of fraud, theft within the system will be substantially reduced and systemic risks eliminated. Systems that were once manual and time consuming can now be automated, saving time, manpower and cutting down costs even further. Possibilities of applying this technology can be seen in tax collection, risk management and conveyancing.

Some applications of the blockchain

Blockchain currently has only a few major applications. However, there are many major institutions that are working on blockchain-based applications. These are aimed at providing flawless services that cut costs, reduce processing times and eliminate fraud and theft.

As a distributed ledger platform for cryptocurrencies. All major cryptocurrencies such as Bitcoin, Dash, NXT and Litecoin all operate on a blockchain.

As a distributed registry such as Factom which provides an unalterable, indelible and secure record keeping system. It uses a distributed and decentralized protocol.

As a decentralized messaging service such as Gems

As a distributed cloud storage service like Sia and Storj

And as a decentralized polling system like Tezos

Developing and future applications of the blockchain

There are a couple of systems and projects being developed and implemented in the various sectors. For instance, the insurance sector has 3 distinct distribution methods which are the micro-insurance, peer-to-peer insurance and parametric insurance. All these are programmed around the blockchain ledger system.

There is online or digital voting which is internet based. This kind of voting system is based on the blockchain and promises simple, accurate and verifiable voting that cannot be compromised. Major banks are still researching on methods they can adopt based on the blockchain. If successful, banks will cut down on back-office operations, reduce transaction processing times and also cut down the cost of operations.

Government and private processes

Book publishing, data storage, land registration and original art identification are all processes that will soon be accomplished via the blockchain. Some of the world's largest financial institutions such as UBS have designed new computer labs where software engineers and researchers can spend more time researching on how the blockchain can be adopted more effectively in the financial services sector.

In Scandinavian countries like Norway and Sweden, progress is well ahead of time in implementing blockchain based programs in the management of land registries. If successfully implemented, it will be very easy to transact in land matters, with parties benefiting from speedy land sale processes. Other European countries are experimenting with property registries based on the same platform.

What are smart contracts?

A smart contract is basically a computer program that can be executed on the blockchain. Smart contracts do not need any human interaction and can be executed automatically either partially or fully. Such contracts can, for instance, execute automated escrows. Escrow is a platform that holds funds from a buyer to be released or paid to a provider upon delivery of certain agreed services. Smart contracts can execute an automated escrow upon the attainment of certain conditions.

Ethereum is a well known platform that functions on the blockchain. One of its main projects, Ethereum Solidity, is a smart contract that can be executed on the platform. It is programmed and implemented using a computer language known as Turin Complete which has the capability of implementing smart contracts.

A great example of the application of smart contracts on the blockchain is found in the music industry. A music tape or CD is produced by a DJ, music producer or even a studio. The tape is then programmed as a smart contract and added to a blockchain. Now each time the music is played, each artist gets paid automatically through a preferred cryptocurrency. A similar smart contract has already been implemented by a US based DJ, Deadly Buda.

Blockchain and the accounting industry

All the four major accounting firms including Ernst and Young have found ways to adopt blockchain in their operations. Take for instance Ernst and Young. All is employees in Switzerland have received cryptocurrency wallets and there is a Bitcoin ATM in their Swiss offices. The other major accounting firms including Deloitte, KPMG and PWC are all working on some form of blockchain which are currently under testing. It is believed that very soon, all these accounting firms will have their own smart applications and will provide more efficient accounting solutions that can be catered and tailored for each individual client, ensuring they received faster, better, more accurate accounting in an efficient and reliable manner.

Benefits of the blockchain

Industries around the world are being completely changed through the blockchain technology. Governments, organizations, payment platforms and financial institutions are all being brought onto the digital platform. Blockchain is revolutionizing how functions are managed and how organizations operate.

Why the blockchain is considered so important ?

In today's world, political, financial, legal and general systems depend on transactions and contracts to function adequately. Transactions and contracts as well as the records kept are vital in the management of assets, data and other resources. In fact, most processes are governed through contracts and associated transactions. The process through which these are recorded and documented is completely outdated. The essential tools, systems and structures needed to record data are not up to date. The blockchain is the platform where all these challenges are solved. Many companies around the world are looking to implement the blockchain technology so they can improve efficiency and enjoy the immense potential benefits therein.

The blockchain eliminates intermediaries like bankers and lawyers

Most of the operations in the modern world require the services of middlemen and other intermediaries. For instance, banks are intermediaries between clients and creditors while lawyers are middlemen between clients and the law. All these middlemen can be eliminated so that members of the general public access the services they need directly. By using blockchain technology, firms, organizations and individuals would be able to interact and deal with each other directly.

This is already happening with bitcoin. Bitcoin is one of the most successful cryptocurrencies in the world. It operates on a decentralized, distributed blockchain. Bitcoin directly connects buyers and sellers, traders and customers, without the need for intermediaries such as banks. Peer-to-peer transactions on the blockchain are faster, cheaper and credible. Transactions are recorded on the blockchain and the information is available for confirmation.

On the blockchain network, each and every transaction and all processes are recorded. It is possible for users to trace back payments or transactions to an organization or user. Each transaction, once confirmed, is verified and has a signature for tracking back. This is the ultimate way of conducting business and undertaking operations with total freedom at very low costs. Transactions will be etched on indelible, tamper-proof, permanent and verifiable system.

Visible advantages of blockchain

Blockchain is a very transparent form of technology. Each user within the system has access to all information. Unlike banks and other organizations which keep information concealed, the blockchain offers transparency and openness that clients appreciate.

The information held on the blockchain is completely tamper-proof. The system is designed to resist hacking and attempts to tamper with the data. Supposing a person wanted to hack information entered in a blockchain with 3000 users. The person would have to hack into all the 3000 computers to change the information. The blockchain is designed to resist alteration to data entered. Should data have to be altered, then it is done by consensus and the process is tedious because it affects other entries.

All transactions on the blockchain are processed instantly. Payments are made and reconciliation done within an instant. Third party systems often take a couple of days to reconcile payments but the blockchain takes a maximum of ten minutes and in some instance, far less time.

There are no fees, charges or transaction costs involved. These are eliminated as all middlemen or intermediaries are made redundant.

Limitations of the blockchain

While the blockchain technology is impressive by any standards, it is not devoid of challenges. The technology used is still relatively new and challenges are encountered that had not been considered earlier.

The blockchain is dynamic. It keeps growing by the day as new blocks are added to it. Scalability therefore becomes an issue because there are challenges regarding amount of storage space required.

The system is still slow at processing transactions. Blockchain can only process 7 transactions per second which is quite low considering VISA can manage up to 56,000 transactions per second.

There is hope that these challenges will be sorted out and it is only a matter of time. There are many software engineers and computer programmers and scientists at work in labs working to improve and perfect the blockchain. It is believed that with time, these challenges will reduce drastically or be eliminated all together.

Chapter 14: Cons Of Cryptocurrency

Lack of Regulation Facilitates Black Market Activity

Probably the biggest drawback and regulatory concern around cryptocurrency is its ability to facilitate illicit activity. Many gray and black market online transactions are denominated in Bitcoin and other cryptocurrencies. For instance, the infamous "dark web" marketplace Silk Road used Bitcoin to facilitate illegal drug purchases and other illicit activities before being shut down in 2014. Cryptocurrencies are also increasingly popular tools for money laundering – funneling illicitly obtained money through a "clean" intermediary to conceal its source.

The same strengths that make cryptocurrencies difficult for governments to seize and track allow criminals to operate with relative ease – though, it should be noted, the founder of Silk Road is now behind bars, thanks to a years-long DEA investigation.

Potential for Tax Evasion in Some Jurisdictions

Since cryptocurrencies aren't regulated by national governments and usually exist outside their direct control, they naturally attract tax evaders. Many small employers pay employees in bitcoin and other cryptocurrencies to avoid liability for payroll taxes and help their workers avoid income tax liability, while online sellers often accept cryptocurrencies to avoid sales and income tax liability.

According to the IRS, the U.S. government applies the same taxation guidelines to all cryptocurrency payments by and to U.S. persons and businesses. However, many countries don't have such policies in place. And the inherent anonymity of cryptocurrency makes some tax law violations, particularly those involving pseudonymous online sellers (as opposed to an employer who puts an employee's real name on a W-2 indicating their bitcoin earnings for the tax year), difficult to track.

Potential for Financial Loss Due to Data Loss

Early cryptocurrency proponents believed that, if properly secured, digital alternative currencies promised to support a decisive shift away from physical cash, which they viewed as imperfect and inherently risky. Assuming a virtually uncrackable source code, impenetrable authentication protocols (keys) and adequate hacking defenses (which Mt. Gox lacked), it's safer to store money in the cloud or even a physical data storage device than in a back pocket or purse.

However, this assumes that cryptocurrency users take proper precautions to avoid data loss. For instance, users who store their private keys on single physical storage devices suffer irreversible financial harm when the device is lost or stolen. Even users who store their data with a single cloud service can face loss if the server is physically damaged or disconnected from the global Internet (a possibility for servers located in countries with tight Internet controls, such as China).

Potential for High Price Volatility and Manipulation

Many cryptocurrencies have relatively few outstanding units concentrated in a handful of individuals' (often the currencies' creators and close associates) hands. These holders effectively control these currencies' supplies, making them susceptible to wild value swings and outright manipulation – similar to thinly traded penny stocks.

Often Can't Be Exchanged for Fiat Currency

Generally, only the most popular cryptocurrencies – those with the highest market capitalization, in dollar terms – have dedicated online exchanges that permit direct exchange for fiat currency. The rest don't have dedicated online exchanges, and thus can't be directly exchanged for fiat currencies. Instead, users have to convert them into more commonly used cryptocurrencies, such as Bitcoin, before fiat currency conversion. This suppresses demand for, and thus the value of, some lesser-used cryptocurrencies.

Limited to No Facility for Chargebacks or Refunds

Although cryptocurrency miners serve as quasi-intermediaries for cryptocurrency transactions, they're not responsible for arbitrating disputes between transacting parties. In fact, the concept of such an arbitrator violates the decentralizing impulse at the heart of modern cryptocurrency philosophy. This means that you have no one to appeal to if you're cheated in a cryptocurrency transaction – for instance, paying upfront for an item you never receive. Though some newer cryptocurrencies attempt to address the chargeback/refund issue, solutions remain incomplete and largely unproven.

By contrast, traditional payment processors such as Visa, MasterCard, and

PayPal often step in to resolve buyer-seller disputes. Their refund, or chargeback, policies are specifically designed to prevent seller fraud.

LIST OF CRYPTOCURRENCY

Cryptocurrency usage has exploded since Bitcoin's release. Though exact active currency numbers fluctuate and individual currencies' values are highly volatile, the overall market value of all active cryptocurrencies is generally trending upward. At any given time, hundreds of cryptocurrencies trade actively.

The cryptocurrencies described here are marked by stable adoption, robust user activity, and relatively high market capitalization (greater than $10 million, in most cases):

Bitcoin

Bitcoin is the world's most widely used cryptocurrency, and is generally credited with bringing the movement into the mainstream. Its market cap and individual unit value consistently dwarf (by a factor of 10 or more) that of the next most popular cryptocurrency. Bitcoin has a programmed supply limit of 21 million Bitcoin.

Bitcoin is increasingly viewed as a legitimate means of exchange. Many well-known companies accept Bitcoin payments, though most partner with an exchange to convert Bitcoin into U.S. dollars before receiving their funds.

Litecoin

Released in 2011, Litecoin uses the same basic structure as Bitcoin. Key differences include a higher programmed supply limit (84 million units) and a shorter target block chain creation time (two-and-a-half minutes). The encryption algorithm is slightly different as well. Litecoin is usually the second- or third-most popular cryptocurrency by market capitalization.

Ripple

Released in 2012, Ripple is noted for a "consensus ledger" system that dramatically speeds up transaction confirmation and block chain creation times – there's no formal target time, but the average is every few seconds. Ripple is also more easily converted than other cryptocurrencies, with an in-house currency exchange that can convert Ripple units into U.S. dollars, yen, euros, and other common currencies.

However, critics have noted that Ripple's network and code are more susceptible to manipulation by sophisticated hackers and may not offer the same anonymity protections as Bitcoin-derived cryptocurrencies.

Ethereum

Launched in 2015, Ethereum makes some noteworthy improvements on Bitcoin's basic architecture. In particular, it utilizes "smart contracts" that enforce the performance of a given transaction, compel parties not to renege on their agreements, and contain mechanisms for refunds should one party violate the agreement. Though "smart contracts" represent an important move toward addressing the lack of chargebacks and refunds in cryptocurrencies, it remains to be seen whether they're enough to solve the problem completely.

Dogecoin

Dogecoin, denoted by its immediately recognizable Shiba Inu mascot, is a variation on Litecoin. It has a shorter block chain creation time (one minute) and a vastly greater number of coins in circulation — the creators' target of 100 billion units mined by July 2015 was met, and there's a supply limit of 5.2 billion units mined every year thereafter, with no known supply limit. Dogecoin is thus notable as an experiment in "inflationary cryptocurrency," and experts are watching it closely to see how its long-term value trajectory differs from that of other cryptocurrencies.

Coinye

Coinye was developed under the original moniker "Coinye West" in 2013, and identified by an unmistakable likeness of hip-hop superstar Kanye West. Shortly before Coinye's release, in early 2014, West's legal team caught wind of the currency's existence and sent its creators a cease-and-desist letter.

To avoid legal action, the creators dropped "West" from the name, changed the logo to a "half man, half fish hybrid" that resembles West (a biting reference to a "South Park" episode that pokes fun at West's massive ego), and released Coinye as planned. Given the hype and ironic humor around its release, the currency attracted a cult following among cryptocurrency enthusiasts. Undaunted, West's legal team filed suit, compelling the creators to sell their holdings and shut down Coinye's website.

Though Coinye's peer-to-peer network remains active and it's still technically possible to mine the currency, person-to-person transfers and mining activity have collapsed to the point that Coinye is basically worthless.

Conclusion

I hope that you now have a better understanding of what Ethereum is and what you can do with it. It isn't as popular as Bitcoin right now but, being a different kind of system, it is fast catching up and is expected, at some point to be a very strong contender, if not take over the top spot for digital currencies.

The next step for you is to do some more research before you take the leap and step in Ethereum. Learn about computer programming if you don't have much knowledge – you will need to know it if you are going to be successful – and join one of the many forums that talk about Ethereum, learn everything you can.

Good luck and I hope that you have a successful future using Ethereum!

www.ingramcontent.com/pod-product-compliance
Lightning Source LLC
LaVergne TN
LVHW011938070526
838202LV00054B/4715